National Center for Construction Education and Research

Gas Pipeline Operations

Prentice Hall
Upper Saddle River, NJ
Columbus, OH

contren™ Learning Series

American Petroleum Institute
Pipeliner Training and Assessment Program (PTAP)

This information is general in nature and intended for training purposes only. Actual performance of activities described in this manual requires compliance with all applicable operating, service, maintenance, and safety procedures under the direction of qualified personnel. References in this manual to patented or proprietary devices do not constitute a recommendation of their use.

---

10 9 8 7 6 5
ISBN 0-13-046671-9

Preface

This volume was developed by the National Center for Construction Education and Research (NCCER) in response to the training needs of the construction, maintenance, and pipeline industries. It is one of many in NCCER's *Contren™ Learning Series*. The program, covering training for close to 40 construction and maintenance areas, and including skills assessments, safety training, and management education, was developed over a period of years by industry and education specialists.

NCCER also maintains a National Registry that provides transcripts, certificates, and wallet cards to individuals who have successfully completed modules of NCCER's *Contren™ Learning Series*, when the training program is delivered by an NCCER Accredited Training Sponsor.

The NCCER is a not-for-profit 501(c)(3) education foundation established in 1995 by the world's largest and most progressive construction companies and national construction associations. It was founded to address the severe workforce shortage facing the industry and to develop a standardized training process and curricula. Today, NCCER is supported by hundreds of leading construction and maintenance companies, manufacturers, and national associations, including the following partnering organizations:

PARTNERING ASSOCIATIONS

- American Fire Sprinkler Association
- American Petroleum Institute
- American Society for Training & Development
- American Welding Society
- Associated Builders & Contractors, Inc.
- Associated General Contractors of America
- Association for Career and Technical Education
- Carolinas AGC, Inc.
- Carolinas Electrical Contractors Association
- Citizens Democracy Corps
- Construction Industry Institute
- Construction Users Roundtable
- Design-Build Institute of America
- Merit Contractors Association of Canada
- Metal Building Manufacturers Association
- National Association of Minority Contractors
- National Association of State Supervisors for Trade and Industrial Education
- National Association of Women in Construction
- National Insulation Association
- National Ready Mixed Concrete Association
- National Utility Contractors Association
- National Vocational Technical Honor Society
- North American Crane Bureau
- Painting & Decorating Contractors of America
- Portland Cement Association
- SkillsUSA-VICA
- Steel Erectors Association of America
- Texas Gulf Coast Chapter ABC
- U.S. Army Corps of Engineers
- University of Florida
- Women Construction Owners & Executives, USA

Some features of NCCER's *Contren™ Learning Series* are:

- An industry-proven record of success
- Curricula developed by the industry for the industry
- National standardization providing portability of learned job skills and educational credits
- Credentials for individuals through NCCER's National Registry
- Compliance with Apprenticeship, Training, Employer, and Labor Services (ATELS) requirements for related classroom training (CFR 29:29)
- Well-illustrated, up-to-date, and practical information

Acknowledgments

This curriculum was revised as a result of the farsightedness and leadership of the following sponsors:

Destin Pipeline/BP
Marathon Ashland Pipeline LLC
MBF Services

This curriculum would not exist were it not for the dedication and unselfish energy of those volunteers who served on the Authoring Team. A sincere thanks is extended to:

Mark Daniels
Nash El-Badawi
David Selvage
Tom Trauthwein

A final note: This book is the result of a collaborative effort involving the production, editorial, and development staff at Prentice-Hall, Inc., and the National Center for Construction Education and Research. Thanks to all of the dedicated people involved in the many stages of this project.

Contents

Module 66101-02

Introduction to the Pipeline Industry

Introduction to the Pipeline Operator Qualification Rule, API, and NCCER

It could be said that petroleum products keep America running. Petroleum products like oil and natural gas are the principal sources of energy in the United States and are used in many manufacturing and chemical processes. Petroleum products are primarily transported by pipeline. There are more than a million miles of liquid and natural gas pipeline in this country, ranging in diameter from two inches to four feet. These pipelines bring crude oil and natural gas from production fields to processing plants and refineries—and ultimately to the consumer.

Pipeline Operator Qualification

Petroleum products are considered hazardous, so pipeline personnel must take extra precautions and follow specific procedures when working with them. To help ensure the safe handling and transportation of petroleum products, the United States Department of Transportation (DOT) mandates safe practices as federal regulations.

One such regulation, the Operator Qualification (OQ) rule, requires pipeline operators to develop a written qualification plan to evaluate their personnel and their contract personnel (see *49 CFR Part 192 Subpart N—Gas*, and *49 CFR Part 195 Subpart G—Liquid*). The evaluation hinges on the operator's ability to perform certain covered tasks and to be able to recognize and respond to abnormal operating conditions that may be encountered while performing these covered tasks.

Covered Tasks

A pipeline performance task is deemed *covered* if the activity passes the following four-part test:

1. Is it performed on a pipeline facility?
2. Is it an operations or maintenance task?
3. Is it performed as a requirement of *49 CFR Part 195 (Liquid)* or *49 CFR Part 192 (Gas)*?
4. Does it affect the operation or integrity of the pipeline facility?

If the answer to all four questions is *yes*, then the activity is considered *covered* and must figure into the evaluation process of pipeline personnel.

American Petroleum Institute

The OQ rule and its related covered tasks provided the catalyst to develop a unified pipeline training and assessment program—so that pipeline operators and owners have ready access to the assessment and training tools they need. This effort was the direct result of the vision and leadership provided by The American Petroleum Institute (API) and API members.

Efforts to unify the industry are nothing new to API. Indeed, the impetus for forming the Institute in 1919 was the need to standardize engineering specifications for drilling and production equipment. Today, the most pressing issues revolve around public perceptions and government policies toward the industry. Speaking with one voice on these issues has become as essential as having interchangeable parts in the field. API's members determine the petroleum industry's positions on public policy as well as the standards that govern its day-to-day operations worldwide.

Consortium on Operator Qualification

API assembled a focus group whose primary purpose was to identify covered tasks and provide compliance recommendations to pipeline operators.

The Consortium on Operator Qualification (COOQ) became the guiding force for the identification and development of covered tasks for the liquid pipeline industry. Their work product, API publication 1161, *Guidance Document for the Qualification of Liquid Pipeline Personnel*, formed the basis for the resulting Pipeliner Training and Assessment Program.

The covered task list created in API publication 1161 has now evolved to include gas covered tasks and is maintained by NCCER. The most current covered task list can be downloaded at www.nccer.org.

National Center for Construction Education and Research

With the guidance and support of API, its members, and the Consortium on Operator Qualification, the National Center for Construction Education and Research (NCCER) was able to design and implement the Pipeliner Training and Assessment Program.

NCCER is a not-for-profit 501(c)(3) education foundation established in 1995 by the world's largest and most progressive construction and maintenance contractors and national contractor associations. NCCER was founded to address the

severe workforce shortage facing the construction and maintenance industries and to develop a standardized training process and curricula. Today, NCCER is supported by hundreds of leading construction and maintenance companies, manufacturers, and national associations and developed more than 35 construction and maintenance titles.

NCCER also maintains a National Registry, which contains the records of trainees and instructors who have received NCCER training. All completion information for every trainee is sent to the NCCER and filed within the Registry. The Registry can then confirm training and skills for workers as they move from company to company, state to state, or even within their own company.

Standardized Curriculum (NCCER's Contren™ Learning Series)

NCCER curricula are developed using the expertise of industry subject matter experts (SMEs). The use of SMEs as an integral part of every new development or revision ensures that the curriculum is current, industry-driven, and indicative of industry practices. The development and/or revision process is facilitated from start to finish by NCCER project managers, who work closely with the SMEs and a select team of technical writer(s) on the development and presentation of the curriculum.

Pipeline: Five Training Paths

The pipeline curriculum focuses on essential skills and competencies within five skill areas:

- Corrosion Control
- Electrical and Instrumentation
- Maintenance
- Mechanical
- Operations (Field and Control Center)

SMEs from the pipeline industry have determined the skills and competencies needed based on API's covered tasks and the consensus of the group/industry. Specific company information is generally not incorporated into the curriculum unless the SMEs determine that the information is the industry standard.

Training Modules and Covered Tasks

The smallest unit of instruction in the curriculum is a *module*. Each curriculum is composed of modules (the equivalent of stand-alone chapters) that are logically sequenced into complete *levels* of instruction. Every module is task-oriented and requires the successful completion of a written (knowledge-based) examination and a performance (hands-on) test. Modules may be individually selected and/or combined for efficient task training. Modules in the pipeline curricula, to the extent possible, will align with API-developed covered tasks. While not all modules will relate to covered tasks, all covered tasks will relate to at least one module.

Benefits of Training to the Pipeline Marketplace

Applying NCCER's standardized training process allows pipeline owners and contractors in the United States to establish the benchmark for acceptable skill and competency levels within the pipeline industry. This process will assist training organizations in developing and/or improving standardized pipeline training programs that effectively match their resources and operating conditions while responding to regulatory demands and to the needs of the industry. The process will also recognize training organizations that meet the minimum standards of competence, warranting regulatory, public, and professional confidence. This in turn will foster national unity within the pipeline industry for the mutual benefit of individual technicians, owners, contractors, and ultimately pipeline-users and society.

Additional Resources

To review the Pipeliner Training and Assessment Program Implementation and Resource Guide and the latest version of pipeline covered tasks, visit NCCER's Web site at www.nccer.org, or contact:

National Center for Construction Education and Research
PO Box 141104
Gainesville, FL 32614-1104
(888) 622-3720
Web site: www.nccer.org

For more information on API, please contact:

American Petroleum Institute
1220 L Street, Northwest
Washington, D.C. 20005
(202) 682-8000
Web site: www.api.org

To order *API Pub 1161*, contact Global Engineering Documents by calling 1-800-854-7149 or visiting their Web site at www.global.his.com.

Additional sources of information on the pipeline operator qualification regulation include:

DOT – Office of Pipeline Safety
www.ops.dot.gov

Transportation Safety Institute
www.tsi.dot.gov

COURSE MAP

This course map shows each curriculum path in the Pipeliner Training program. The suggested training order begins on the left with the Pipeline Core modules and proceeds to the right. Which curriculum path you follow depends on your areas of specialty and interest. On a given curriculum path, skill levels increase as you advance to the right on the course map, although some are single-level curricula. The local Training Program Sponsor may adjust the training order.

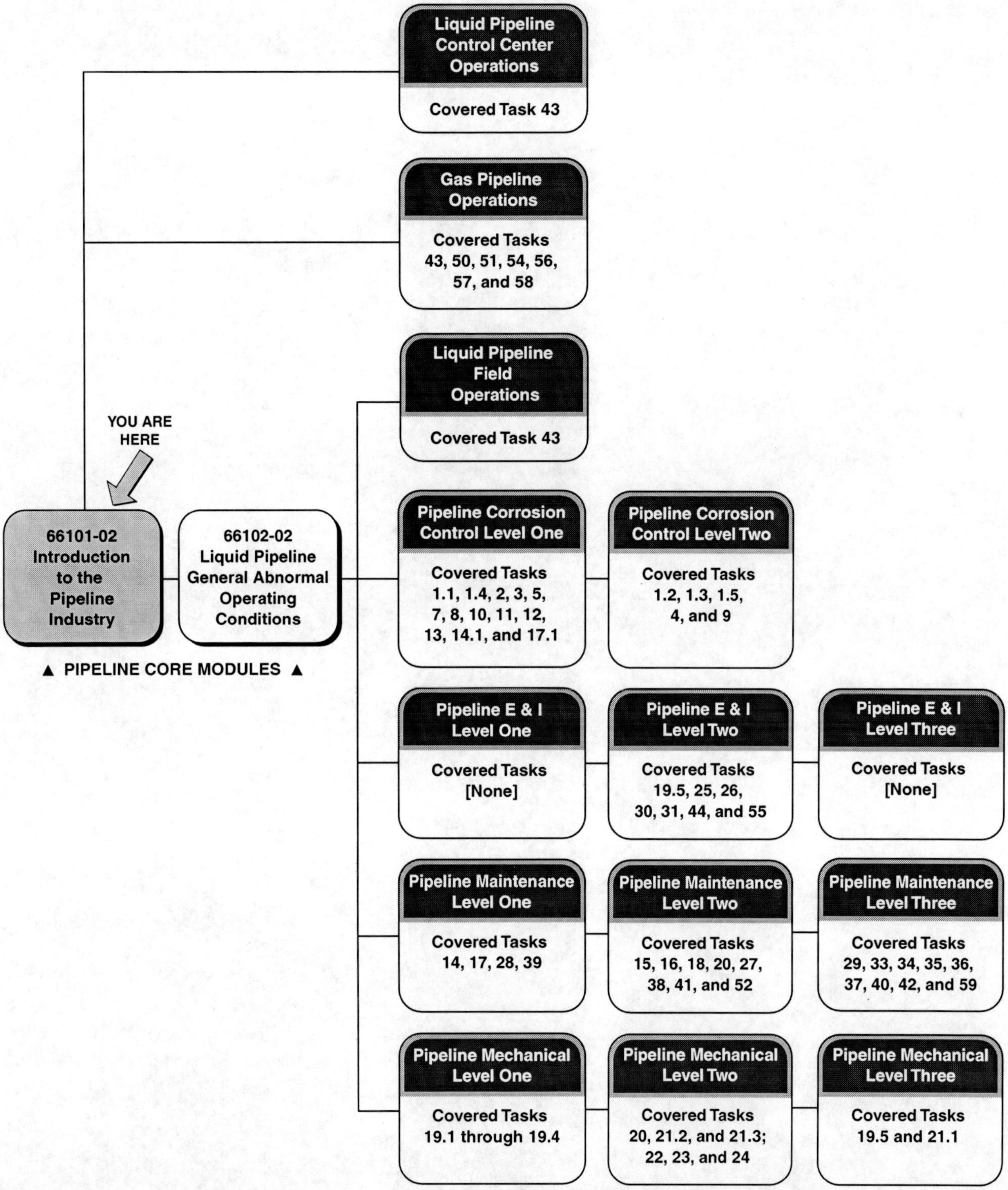

101CMAP.EPS

MODULE 66101-02 CONTENTS

MODULE 66101-02 CONTENTS (Continued)

Figures

Tables

MODULE 66101-02

Introduction to the Pipeline Industry

Objectives

When you have completed this module, you will be able to do the following:

1. Explain the basic functions and purposes of pipelines and facilities and identify the characteristics and hazards of common pipeline products.
2. Identify maps and drawings used to depict pipelines and facilities.
3. Explain the roles of control personnel and equipment in the overall operation of a pipeline.
4. Explain liquid pipeline hydraulics and gas pipeline pneumatics.
5. Explain the types and purposes of pipeline equipment.
6. Explain pipeline electrical power systems and corrosion control.
7. Review operations, maintenance, and emergency procedures and perform documentation required for pipeline operations.

Prerequisites

Before you begin this module, it is recommended that you successfully complete the following: Core Curriculum.

1.0.0 ◆ INTRODUCTION

Pipelines are used to transport **petroleum**, petroleum **products**, liquid carbon dioxide, and other liquids and gases over great distances. In spite of the costs of constructing pipelines, they are a relatively economical means of transport. A pipeline system consists of pipes, pump stations, compressor stations, and other facilities to support the safe, continuous movement of liquids and gases. Pipelines carry petroleum from the source, the oil and gas fields, to the **refineries**, to and from gas processing plants, and eventually to consumers. Pipes are most often buried in the ground, but they are also laid under water and sometimes erected above the ground.

Today, a network of gas and liquid pipelines crisscrosses the United States. The first pipeline on a continental scale was completed in the 1930s, but the real boom in cross-country lines came during World War II when the Big Inch and the Little Big Inch were constructed to provide fuel for wartime industry and the military. The Big Inch began operating in 1943, was 24 inches in diameter, and stretched 1,341 miles from Longview, Texas, to Phoenixville, Pennsylvania, and Linden, New Jersey. This made it the largest **crude oil** pipeline in the world. The Little Inch, opened in 1944, was 20 inches in diameter and 1,475 miles long. It carried **refined products** from Beaumont, Texas, to Linden, New Jersey.

The most famous crude oil pipeline in the U.S. crosses 800 miles of permanently frozen ground in Alaska. The Trans-Alaska Pipeline, which began operations in 1977, is 48 inches in diameter and carries crude oil from the Prudhoe Bay field to the shipping terminal at Valdez, Alaska. Along the way, it crosses 34 major rivers and streams and three major mountain ranges, going from sea level to as high as 4,800 feet and back to sea level. Protection of the delicate tundra, permafrost, wilderness environment, and wildlife in its path were and still are critical concerns.

All pipeline operators are concerned with safety and environmental impact, particularly those carrying petroleum products that are hazardous to the environment as well as to the people who live near the pipeline. Good construction practices, good maintenance and operations practices, continuous monitoring, and rapid response are necessary to prevent or to minimize the effects of pipeline leaks. Operations of a pipeline must be conducted in a fashion that will ensure the safety of employees and the general public and protect the environment.

The petroleum pipeline system can be divided into the following functions:

- Origination
- Transport
- Storage
- Delivery

1.1.0 Origination

All petroleum products produced or shipped originate from the domestic and foreign crude oil or natural gas from the oil and gas fields. Exploration teams from petroleum companies use geological and geophysical methods to find possible deposits of oil and gas deep under the layers of the earth. Drilling confirms and accesses the find; the well is completed and production begins.

1.1.1 Exploration

Oil and gas are formed in the earth from decaying organic matter that was buried long ago. The combination of time, pressure, and heat causes the matter to undergo a chemical change into hydrocarbons. These fluids and gases then tend to move, or migrate, upward in the earth until they either escape at the surface or are trapped in the earth's layers. It is this trapped fluid and gas that is sought by exploration.

The oil/gas-bearing formation is basically the layer of earth in which the hydrocarbon deposit is trapped until it is drilled. The formation must have the porosity to hold the fluid and gas and have the structure to keep it from moving further upward. Traps can have a variety of structures, and some of the most typical are the anticline trap, stratigraphic trap, fault trap, and salt dome trap, as shown in *Figure 1*.

1.1.2 Drilling and Completion

When a possible deposit is indicated, a well must be drilled to try to find it. Many wells have been drilled without finding the sought-after oil and gas. There may not be enough to warrant production costs or it cannot be drawn out economically. However, when a workable deposit is found, the well is completed for production. Completion includes casing the well with concrete, perforating the casing with

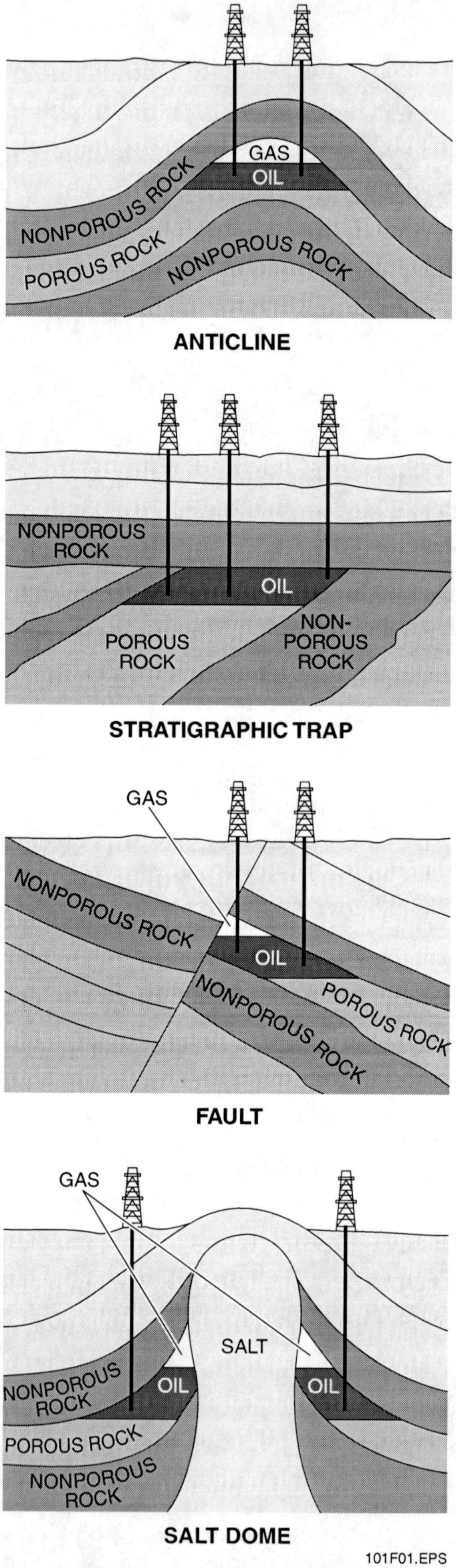

Figure 1 ◆ Types of traps.

holes for the product to flow through tubing within the casing, and installing pumps or valves on the top of the hole to control the flow. Tests of the producing well may show a large enough deposit to warrant other wells being drilled and completed in the area. All of these wells comprise a field. It is from these wells that the oil or gas is gathered to transport through the pipeline.

Many crude oil reservoirs have a gas cap. The oil produced from these reservoirs often contains various amounts of associated gas. Some formations, however, trap only gas. This trapped gas forms a gas or gas-condensate reservoir, which produces only gas or gas with varying amounts of liquid gas condensate. Drilling gas wells is very similar to drilling and completing crude oil wells. The primary difference between crude oil and gas wells is in the type of surface facility equipment needed to process the oil or gas after it is produced.

1.1.3 Production

Different types of pumps or valves, such as a Christmas tree and a beam pump, are used to cap the well depending on the properties of the product being withdrawn and the formation producing it (*Figure 2*). Oil and gas are often found in the same formation along with some amount of water. The amount of each may vary in relation to each other, and this may affect the production of the well. Since gas is lighter than oil, and oil is lighter than water, they are layered in that order: gas on the top, then oil, then water on the bottom.

Because the formation had to be closed off in some way for the product to accumulate, pressure is usually also trapped in the formation. When it is drilled, that pressure can push the oil, gas, and water into the hole with some force. When it is controlled, this pressure is good because it can push the oil and gas up to the surface with little or no pumping. Generally a newer well will have more pressure than an older one, so in time, pumping power may be needed to pull the product out. However, some wells start with only a little pressure and must be pumped from the beginning. Control of the flow is important since producing a well too fast can affect the life of the well. These conditions, along with other properties, will determine how the well is capped and produced. A Christmas tree of valves may be installed to control the flow, and a beam pump may be used to pull the product out. Gas wells are completed with only a Christmas tree.

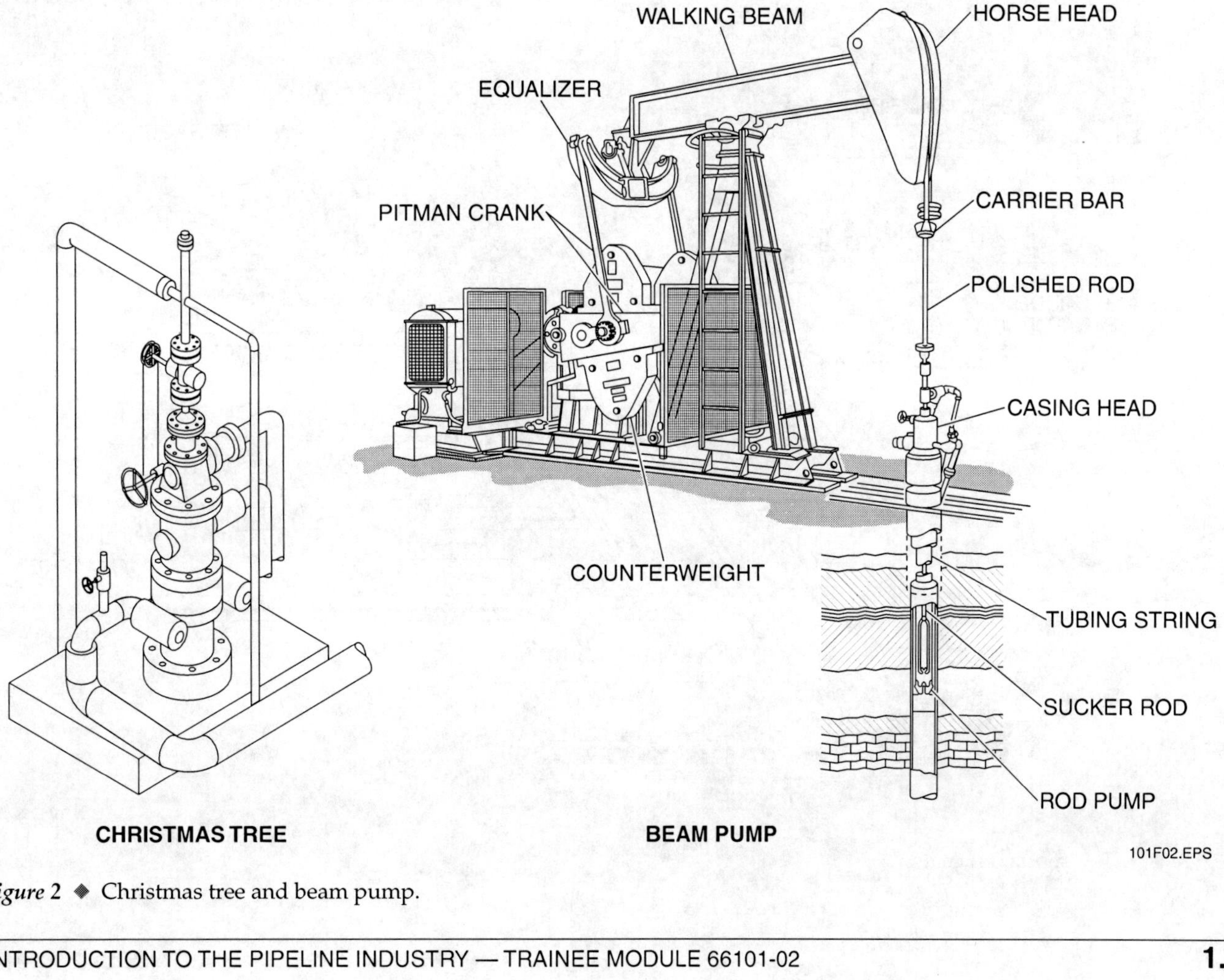

Figure 2 ◆ Christmas tree and beam pump.

1.2.0 Transport

Lines from the oil and gas fields are called *gathering lines* or *flow lines*. These gathering lines bring the oil and gas together into a facility for separation and processing. The vapor gas must travel in a different pipeline from the crude oil, so the gas and oil are separated in huge high-pressure **tanks**, called *gas/oil separators*. *Figure 3* shows the high-pressure separators.

From the separation or gas processing facility, the crude oil and gas is directed into a large **main line**, or trunk line, for the longest part of the journey to a refinery or a shipping terminal. The gas may be stored in tanks or routed through gas pipelines to market. Gas main lines transport gas to a gas distribution center for distribution to consumers, industrial facilities, or storage caverns. From the refinery, a large main line, or trunk line, will carry products across country with distribution lines branching off the main line along the way. A large distribution line may carry the liquid to a central distribution point, such as a regional or city station, from which smaller lines or vehicles carry it to consumers. Pipelines are constructed to meet *ASME Piping Standard, B31.4-1998, Pipeline Transportation Systems for Liquid Hydrocarbons, and other Liquids*. *Figure 4* shows the transportation chain.

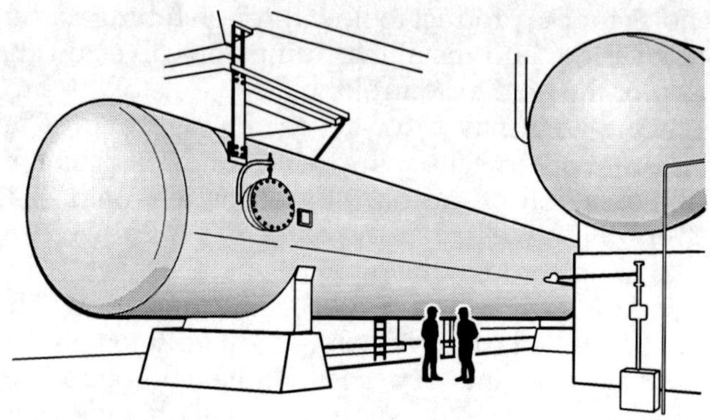

Figure 3 ◆ High-pressure separators. 101F03.EPS

1.2.1 Pipe

The size of pipe used in pipelines varies from 2-inch diameter pipe for gathering lines or small distribution lines, to 48-inch pipe for main lines or trunk lines. Most modern pipe is made of steel, either seamless or with a welded longitudinal seam, that is produced in lengths that are 40-60 feet long. These are often welded into 80-foot sections before being delivered to the construction site. Steel pipe is vulnerable to **corrosion**, which can cause damage to the pipe and reduce wall thickness, compromising the integrity of the pipe. Pipe coating and wrapping are the primary

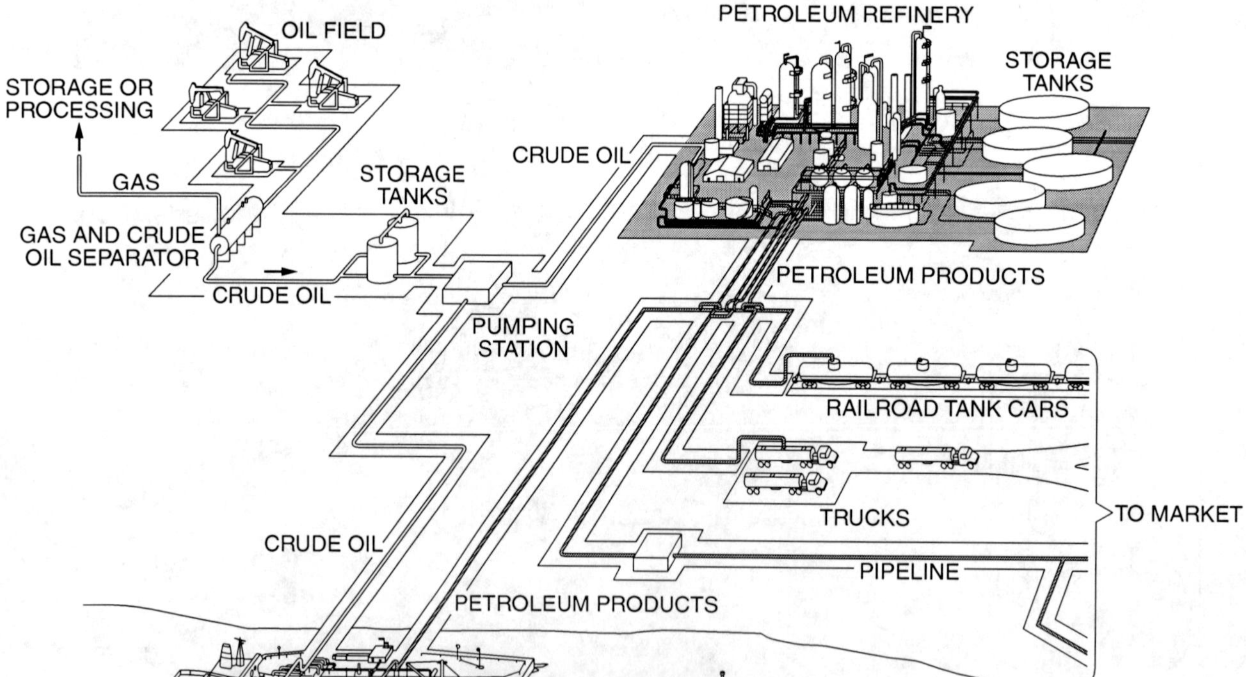

101F04.EPS

Figure 4 ◆ Transportation chain.

means of protecting the pipe. **Cathodic protection** is used after the pipe is buried to inhibit a corrosive environment.

1.2.2 Construction

At the time of construction, permits are obtained for a specified right-of-way for the pipeline. Earth-moving equipment clears the land, an access road is built along the route, and trenches are excavated for the pipe. Trenches require a minimum depth of coverage as stated in *Pipeline Safety Regulations, Part 195.248* for liquid and *Part 192.327* for gas.

Pipe is hauled to the right-of-way and welded in strings by teams of welders. After welding, radiographic inspection is used to check the welds. Sometimes, pipe must be bent by bending machines to fit the terrain or to go around obstacles. To inhibit corrosion, the pipe must be coated and wrapped before burial. This can be done at the mill or at the site.

The trench is dug and then prepared for the pipe with a layer of protective bedding. The pipe is strung or laid in the trench by sideboom tractors. Long sections are tied-in or welded together to form a continuous conduit. The trench is then filled with backfill to cover the pipe (see *Figure 5*). The pipe must be **hydrostatically tested** or checked for leaks. This involves filling the line with water and pressurizing it to see if the system will maintain the pressure intended.

When piping crosses lakes and rivers, it may be laid and weighted with concrete or steel anchors to keep it on the bottom. *Figure 6* shows an add-on weight installation. In most rivers, the crossings are drilled under the river bed without disturbing the bottom.

Pipelines are usually buried except at pump and compressor stations, some valve stations, and terminals. However, pipe is sometimes erected aboveground by using H-shaped supports, called bents, to hold the pipe. These bents are located at specified intervals along the line. The uprights on the bents are vertical support members (VSMs). In areas like Alaska, the VSMs have to be self-cooling to avoid heating the permafrost. This is not the normal installation, but it is necessary in frozen environments. *Figure 7* shows a bent with self-cooling VSMs.

Supports have sliding plates, or shoes, to allow the pipe to contract or expand with weather or to move with **seismic activity**. The shoes allow the pipe to move as much as 20 feet back and forth and 10 feet up and down. Aboveground lines are also constructed in a zigzag pattern instead of a straight line to allow for movement. *Figure 8* shows a common elevated pipeline section.

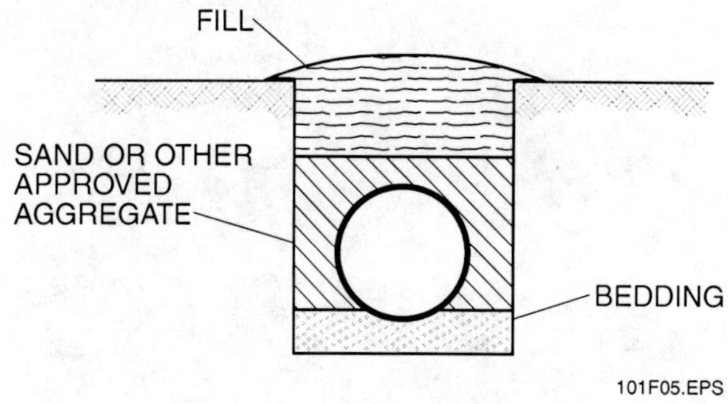

Figure 5 ◆ Pipe burial.

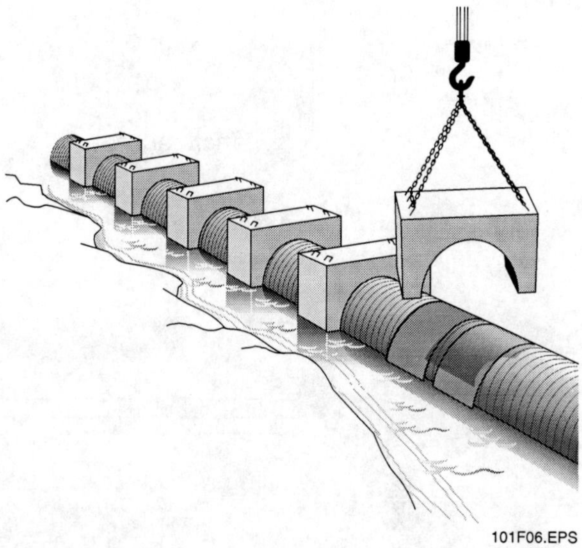

Figure 6 ◆ Add-on weight installation.

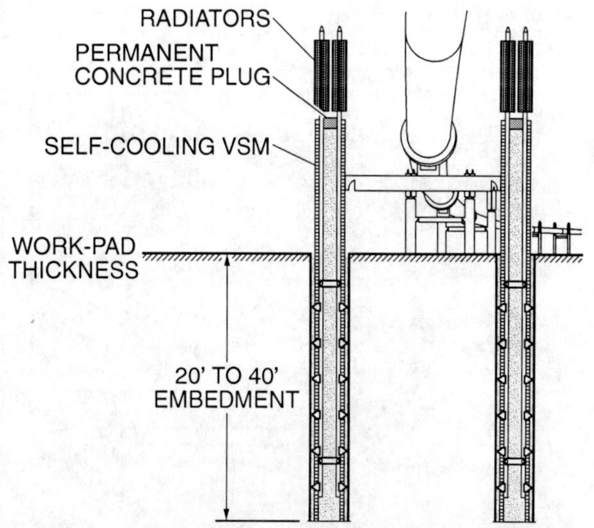

Figure 7 ◆ Bent with self-cooling VSMs.

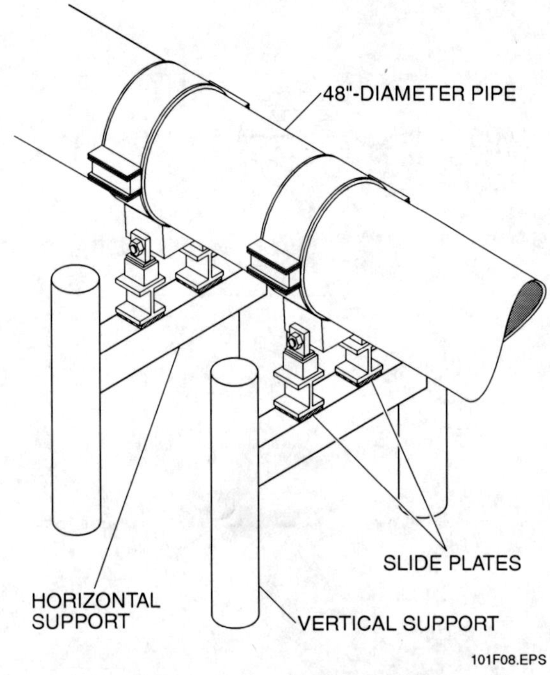

Figure 8 ◆ Common elevated pipeline section.

Pump and compressor stations are spaced strategically along the pipeline at intervals that are dependent on ground elevation, terrain, and other pipeline design factors to maintain the desired flow rate in the line. The pressure that causes the fluid or gas to flow is lost mostly through friction of the product but also by changes in elevation. This pressure must be changed by pumps and compressors to maintain flow. Drag-reducing agents (DRA) can also be added to the product in the line to reduce the friction loss.

A pump or compressor station facility includes pumps or compressors for moving great volumes of liquid or gas, methods of controlling the flow and pressure, means of measuring and accounting for the commodities being moved, storage facilities, and other devices for quality control. *Figure 9* shows an example of a liquid pipeline pump station layout. *Figure 10* shows an example of a gas pipeline compressor station layout.

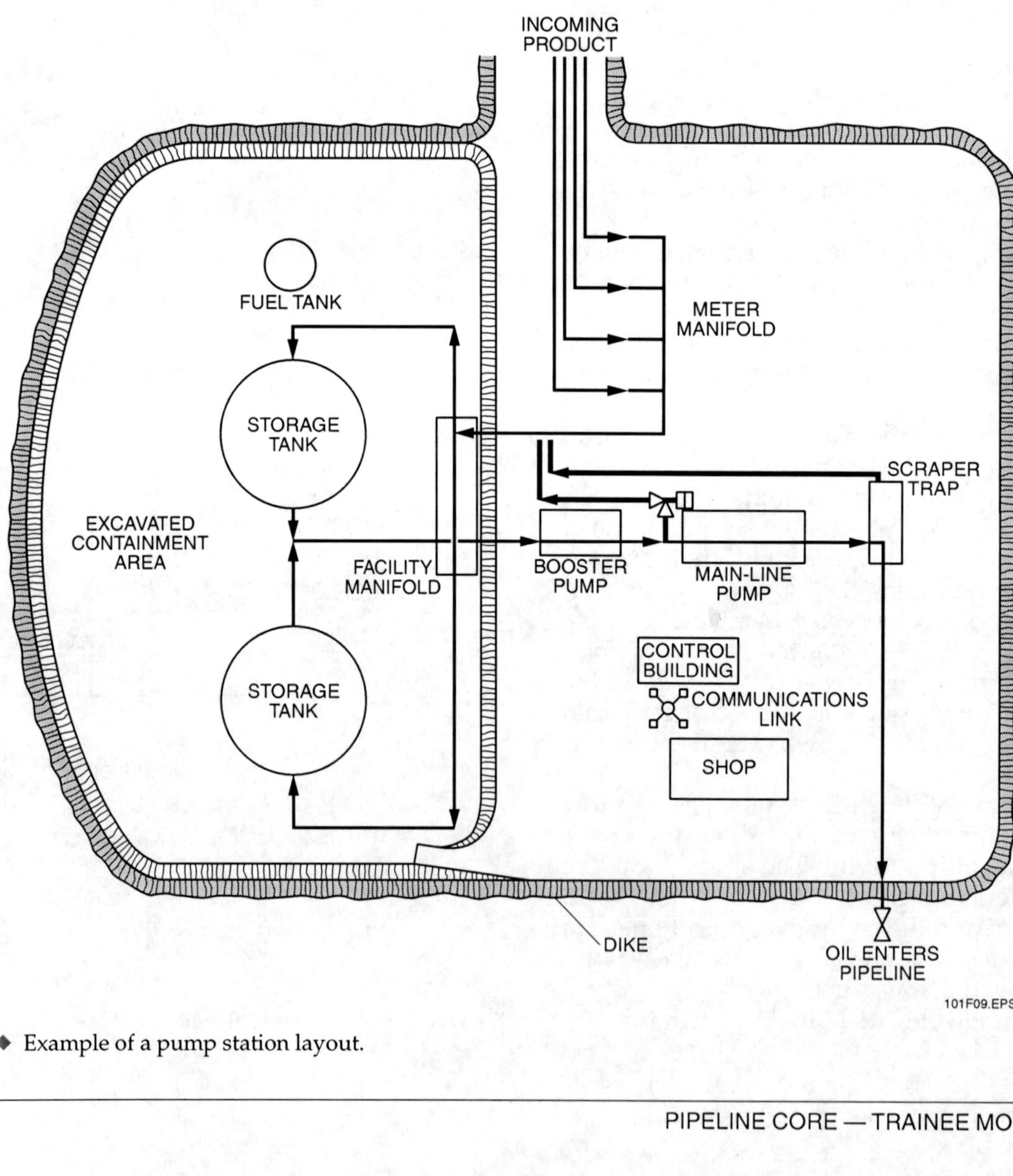

Figure 9 ◆ Example of a pump station layout.

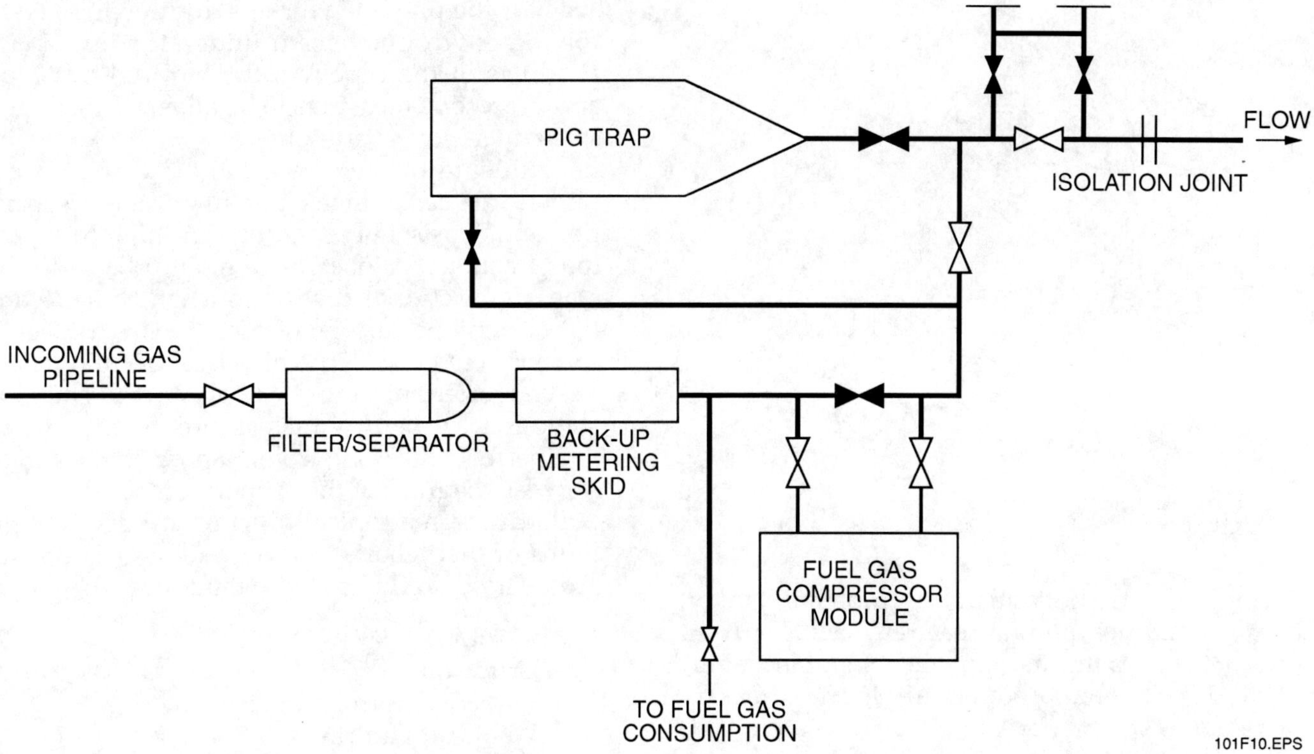

Figure 10 ◆ Example of a gas pipeline compressor station layout.

1.2.3 Operation and Maintenance of the Pipeline

The pipeline is monitored closely and inspected for potential equipment failure and line integrity. Crews identify problems and dig up the pipe to perform repairs. Small **anomalies** may be repaired temporarily while large anomalies may require replacing the pipe. In this case, a bypass line may be constructed so that flow continues while the damaged line is repaired.

The inside of the pipe must also be cleaned periodically to remove wax, scale, and debris. A scraper **pig** is inserted into the line at one point and flows with the liquid or gas to another point, where it is removed. The pig scrapes the inside walls using the hydraulic force of the flow behind it. An instrument pig, also called a smart pig, is inserted into the line periodically to check for anomalies.

1.2.4 Petroleum Products

Different petroleum products can be shipped consecutively from the point of origin in the same pipeline. This is called batching. To maintain quality, batches of each product are arranged in the line and pumped under pressure so that little or no mixing occurs during transit.

1.3.0 Storage

Storage facilities may be located along the pipeline route, at the origination sites, at pump stations, and at terminals. Storage is necessary to meet seasonal needs and other demand fluctuations. Storage is also necessary for the efficient operation of production equipment, pipelines, tankers, refineries, and for emergency use. Large volumes of natural gas are typically stored in underground storage fields throughout the United States.

The trend towards giant tankers created a need for large storage facilities at the loading and unloading ends of tanker runs. A large crude oil terminal typically has several tanks, each with capacities up to half a million **barrels**. A small pump station might have only one tank with a capacity of about 50,000 barrels. Stations along a product line may need separate tanks for storage of each product coming through. These are sometimes called relief tanks or surge tanks. *Figure 11* shows some storage tanks.

1.4.0 Delivery

A wide range of delivery methods is used for gas and petroleum products depending on their destination, intended use, and transport properties.

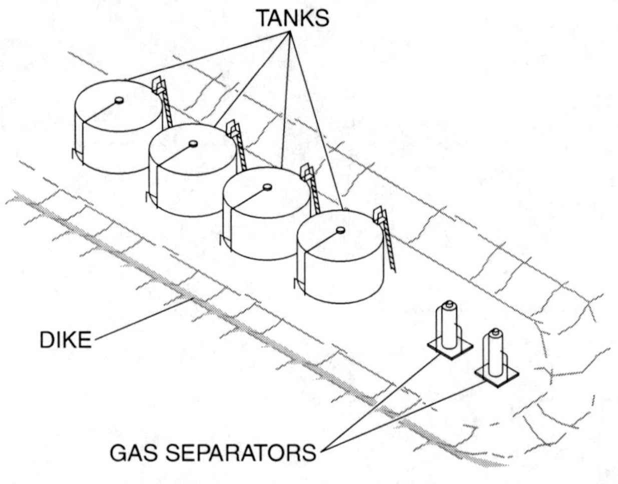

Figure 11 ◆ Storage tanks.

Petroleum is carried chiefly by pipeline, tanker, barge, tank truck, and railroad tank car. Nearly all product moves through pipelines for at least part of the route. Some products are delivered to city or regional stations by pipeline. From there they may be piped or delivered to local consumers or distribution sites. Gasoline, diesel, and kerosene are delivered by tank truck to gas stations for purchase by users. Industrial products may be shipped to companies by pipeline, truck, or railroad. Supertankers are used to move product on ocean routes, while barges are used on rivers and canals. While liquid petroleum products are typically measured in barrels, gas is typically measured using cubic feet. Depending on the volume being measured, gas quantities can be measured using billion cubic feet (BCF), thousand cubic feet (MCF), or hundred cubic feet (CCF). *Table 1* shows typical capacities that can be moved by different means of transport.

Table 1 Means of Transport

Transport	Capacity
Tanker Truck	200 barrels
Railroad Tanker	1,500 barrels
Barge	15,000 barrels
Supertanker	1 million barrels+
Large Liquid Pipeline	2 million barrels+/day
Large Gas Pipeline	1 BCF/day

101T01.EPS

2.0.0 ◆ PIPELINE PRODUCTS

A pipeline can carry a wide range of petroleum and gas products. Each of these products has its own characteristics and rules for handling. Each of these products also presents its own set of hazards for the pipeline workers, for the public, and for the environment. An understanding of the products in the pipeline, the hazards involved, and protective measures to be taken is essential to safe and secure operations.

Petroleum products are produced from crude oil in three basic stages. The first stage is separation and consists of separating the oil into its various fractions, or parts. The main process in this stage is fractional distillation, which separates light, medium, and heavy fractions. The second stage is conversion, in which less useful fractions are converted into more valuable ones. The third stage is treatment, which improves the quality and performance of petroleum products. *Figure 12* shows a diagram of the refinery process.

Products in a typical pipeline are discussed in terms of their characteristics and associated hazards. These products may include:

- Natural hydrocarbons
  - Crude oil
  - Liquefied petroleum gas
  - Wellhead natural gas
- Refined products
  - Gasoline
  - Diesel fuel
  - Fuel oil
  - Jet fuel
  - Consumer-ready natural gas
- Special products
  - Chlorine gas
  - Hydrogen chloride gas
  - Propylene
  - Acetone
  - Benzene
  - Ethane and ethylene
  - Anhydrous ammonia
  - Methane gas
  - Carbon dioxide

2.1.0 Natural Hydrocarbons

Natural hydrocarbons include crude oil, liquefied petroleum gas, and natural gas.

2.1.1 Crude Oil

Crude oil is unrefined petroleum—the source of the myriad of petroleum products. Its specific gravity ranges from 9 to 55 as compared with the specific gravity of water, which is 10. A light crude is one with a higher gravity, and a heavy crude has a lower gravity.

Because the petroleum may have different ingredients or fractions, its color can range from yellow

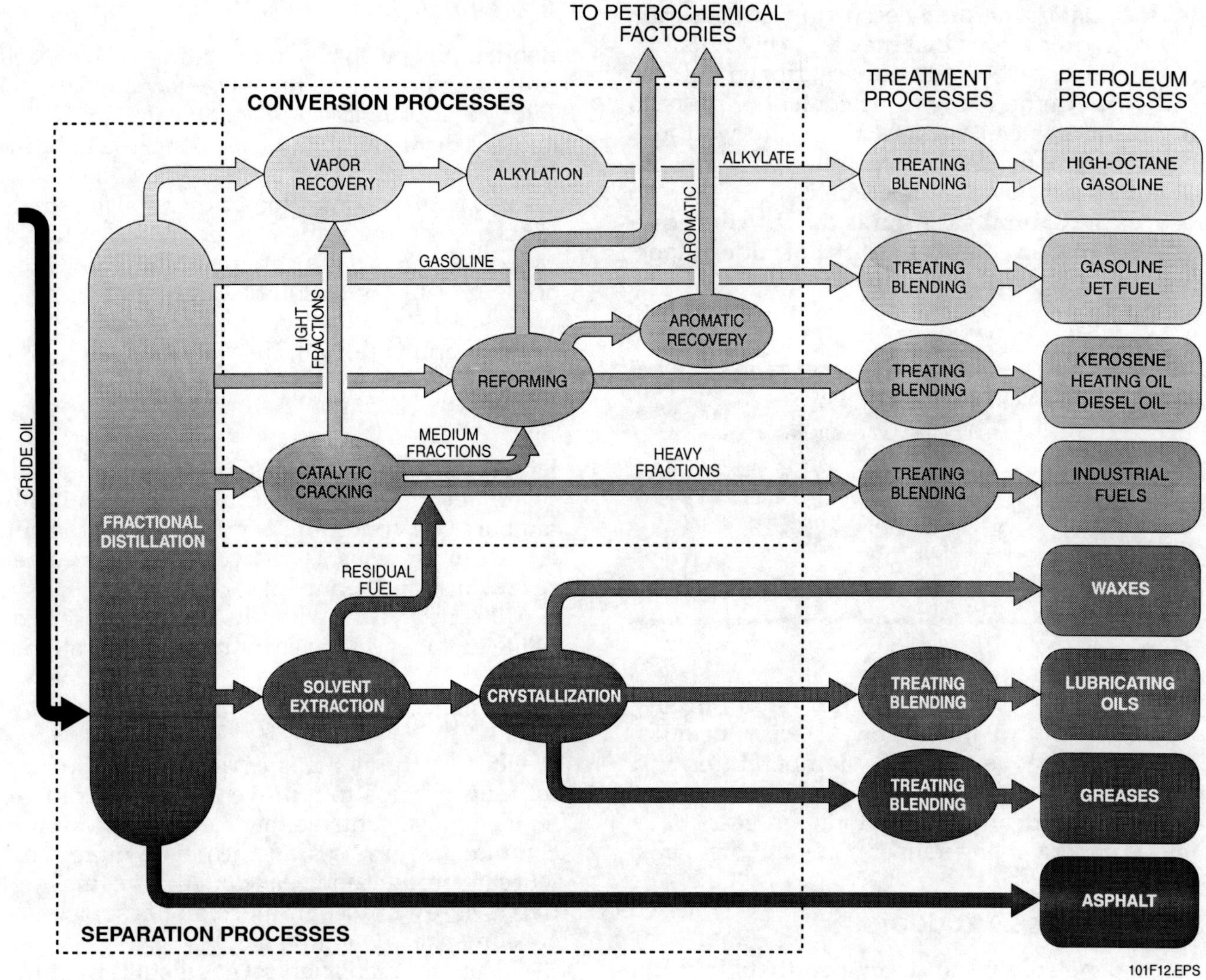

Figure 12 ◆ Diagram of the refinery process.

to black. It may have a paraffin, asphalt, or mixed base. Crude oil also contains substances other than hydrocarbons. These impurities include metallic compounds and sulfur and may make up as much as 10% of the oil. Oil containing a large amount of sulfur compound is called a sour crude. A sweet crude contains little or no sulfur.

2.1.2 Liquefied Petroleum Gas

Liquefied petroleum gas (LPG) is a mixture of heavier, gaseous, paraffinic hydrocarbons, principally butane and propane. These gases, easily liquefied at moderate pressures, may be transported as liquids and converted to gases on release of the pressure. Thus, LPG is a portable source of thermal energy that finds wide application in areas in which it is impractical to distribute natural gas. LPG is used in industry for cutting and welding metals and on farms for operating various kinds of machinery. The principal sources are natural and refinery gas, from which the liquefied petroleum gases are separated by fractionation.

Butane/butylene mix, a form of LPG, is extremely flammable. It releases vapors at well below ambient temperatures and readily forms flammable mixtures with air. If ignited, it will burn in open air or explode in confined spaces. Its vapors are heavier than air and may travel long distances to a point of ignition and then flash back. Alkane/chlorine gas mixtures have produced explosions. Closed systems will form a white frost at the point of the release. Liquid spills will vaporize, forming cold, dense vapor clouds that should be avoided even with respiratory equipment.

2.1.3 Wellhead Natural Gas

Natural gas is the name given to the naturally occurring hydrocarbon mixtures found in many underground wells. Natural gas is an unrefined product that is a naturally occurring gaseous mixture of hydrocarbons that may also contain nitrogen, helium, carbon dioxide, hydrogen sulfide, and water vapor. Because the content or makeup of natural gas can vary from well to well and region to region, it is normally referred to as wellhead natural gas. Natural gas is processed to recover the **natural gas liquids (NGLs)**, the heavier hydrocarbon components that include butane, propane, and natural gasoline.

> **WARNING!**
> Exposure to petroleum products can cause eye, skin, and respiratory tract irritation. Always wear proper personal protective equipment (PPE) and refer to company procedures and the product material safety data sheet (MSDS) for specific details.

Natural gas is highly compressible, highly expandable, and has a low specific gravity. Although gaseous at normal temperatures and pressures, the gases making up the mixture are variable in form and may be found as gases or as liquids under varying temperatures and pressures.

2.2.0 Refined Products

Refined petroleum and gas products include gasoline, jet fuel, diesel fuel, heating oil, and kerosene. Other refined products may include propane, butane, consumer-ready natural gas, and blending components.

2.2.1 Gasoline

Gasoline is a volatile, flammable liquid hydrocarbon refined from crude oil and used universally as a fuel for internal-combustion, spark-ignition engines. Natural gasoline, or casinghead gasoline, is the liquid hydrocarbon recovered from wet natural gas. Gasoline is classified as regular, premium, and aviation grades according to how smoothly it burns in an engine.

2.2.2 Diesel Fuel

Diesel fuel is a light hydrocarbon mixture for diesel engines, similar to furnace fuel oil. It has a boiling range just above that of kerosene. Diesel fuel requires less refining than gasoline.

2.2.3 Fuel Oil

Refined fuel oil comes in two grades. The specifications for both are defined by *American Society for Testing and Materials (ASTM) Specification D 396*.

No. 1 distillate, such as kerosene, is a light fuel. No. 2 fuel oil diesel is a distillate fuel oil prepared for use in atomizing-type burners or for smaller industrial burner units. Lesser grades of fuel oil, called *residual fuel oils*, are made from the topped crude of refinery operations. Residual fuel oil is used for the production of electric power, space heating, and various industrial purposes.

2.2.4 Jet Fuel

Jet fuel is a highly refined kerosene petroleum distillate. Jet fuel has good flow properties in low-temperature conditions, and it burns efficiently and cleanly. It must be delivered pure and free of rust, water, and sediment.

Military jet fuel is a highly refined kerosene distillate that meets stricter military specifications.

2.2.5 Consumer-Ready Natural Gas

Wellhead natural gas is normally run through a refining process to remove any undesired contents, remove contents that can be used for other commercial purposes, and achieve a more consistent safe product and heat content. The final product is known as commercial natural gas or consumer-ready natural gas.

Consumer-ready natural gas is still a mixture of different gases but is made up primarily of methane. Other gases, such as ethane, propane, nitrogen, and carbon dioxide can also be found in trace quantities. The composition of the consumer-ready natural gas remains fairly consistent over time. The heat value range of typical consumer-ready natural gas is 1,020 to 1,045 Btus per cubic foot.

2.3.0 Special Products

Special products include other by-products that are produced and carried through pipelines.

2.3.1 Chlorine Gas

Chlorine gas is a product used for water purification, chemical processes, and the manufacture of plastic. Chlorine gas is a strong oxidizing agent; it can react explosively with organic products and it is poisonous. Liquid and gas chlorine under pressure can be hazardous. Chlorine is nonflammable and nonexplosive, but like oxygen, it can support combustion of certain substances.

It can react violently or form explosive compounds with many chemicals, such as acetylene, turpentine, ether, ammonia gas, and hydrogen, and finely divided metals.

2.3.2 Hydrogen Chloride Gas

Hydrogen chloride gas is a fuming, poisonous gas used to make hydrochloric acid, an important industrial chemical used in many chemical compounds, in metallurgy, in food processing, and as a cleaning agent. Hydrogen chloride is nonflammable and noncombustible. However, if dissolved in water, it produces hydrochloric acid, which attacks most metals.

2.3.3 Hydrogen Sulfide

Hydrogen sulfide is a flammable, colorless, gaseous compound of hydrogen and sulfur, which in small amounts has the odor of rotten eggs. It is formed naturally by the decomposition of organic animal and vegetable matter by bacterial action. Hydrogen sulfide is found in crude oil, natural gas, sewers, and swamps and is created by a number of industrial processes. It is extremely corrosive and poisonous. Hydrogen sulfide causes damage to the eyes, skin, breathing passages, and lungs, attacks and paralyzes the central nervous system, especially the parts that control the lungs and heart. In large amounts, it deadens the sense of smell. Hydrogen sulfide is an extremely dangerous compound, and exposure is deadly.

2.3.4 Propylene

Propylene is used in plastics manufacturing. It is extremely flammable and will readily ignite at ambient temperatures. The flammable vapors, being heavier than air, may travel long distances along the ground to a source of ignition and then flash back. Propylene presents the danger of rocketing cylinders, ruptured containers, and explosion. Reignition or explosion can occur if the flames are extinguished without stopping the source flow. An electrostatic charge may build up during handling if the equipment is not grounded.

2.3.5 Acetone

Acetone is a clear, colorless liquid with a sweet, pungent odor that is used as a solvent. It is extremely flammable and may cause flash fire. It is incompatible with a variety of materials, including inorganic acids, caustic, amines, alkanolamines, oxygen, halogens, aldehydes, ammonia, oxidizing agents, and chlorinated compounds. Contact with these should be avoided during transport, handling, and storage.

2.3.6 Benzene

Benzene is an extremely flammable liquid, readily igniting at ambient temperatures. The vapor forms an explosive mixture with air. Flammable vapors, being heavier than air, may travel long distances to a point of ignition and flash back. Incomplete combustion can produce carbon monoxide.

2.3.7 Ethane and Ethylene

Ethane and ethylene are simple asphyxiants. They are dangerous when exposed to heat or flame and can react violently with oxidizing materials. Both are flammable over a wide range in air, with an even broader range in pure oxygen.

2.3.8 Anhydrous Ammonia

Anhydrous ammonia is the liquid form of pure ammonia gas. It is used as a fertilizer because of its high nitrogen content. It is also used as an industrial refrigerant in cold-storage and ice manufacturing plants, since it absorbs a large amount of heat when going from a liquid back to a gas. Anhydrous ammonia is made by compressing dry ammonia gas.

Ammonia is a colorless alkaline gas made up of one part nitrogen and three parts hydrogen. It is lighter than air and has a sharp, stinging odor. Ammonia can be safely inhaled if it is greatly diluted in air, but concentrated ammonia gas can cause suffocation and death. It does not burn in air, but it burns in oxygen with a weak yellow flame. Ammonia is made by combining hydrogen and nitrogen, and as a by-product in the production of coal and coke gas.

2.3.9 Methane Gas

Methane is a member of the alkaline or paraffin series of hydrocarbons. It is called marsh gas because it forms by anaerobic bacterial decomposition of vegetable matter in swampy land. Coal miners know it as *firedamp* because it is combustible when mixed with air. It is a major component of natural gas and of coal gas, and forms in large amounts in sewage disposal processes.

In addition to its use as a fuel, methane is important as a source of organic chemicals and of hydrogen. In the presence of steam and catalysts, it yields carbon monoxide and hydrogen, which can be converted to liquid hydrocarbons or to methanol and other alcohols. The incomplete combustion of methane with air produces carbon black, which is used in rubber and ink. Chlorination of methane yields various chlorides and chloroform.

2.3.10 Carbon Dioxide

Carbon dioxide (CO_2) is a colorless, odorless, tasteless gas which is about 15 times heavier than air. Under normal conditions, it is stable, inert, and nontoxic. The decay of all organic matter and the burning of hydrocarbon fuels produces carbon dioxide. It is obtained commercially from four sources: naturally occurring underground CO_2 wells, fermentation, combustion of carbonaceous fuels, and as a by-product of chemical processing.

Carbon dioxide gas may be liquid or solid based on compression and lowered temperatures. It may be used as a refrigerant, inerting medium, chemical reactant, neutralizing agent for alkalies, and pressurizing agent. Liquid carbon dioxide is an effective fire-extinguishing agent because it rapidly reduces the temperature of burning materials below the ignition point. The gas, being heavy and inert, displaces air, which is the source of the fire's oxygen, and blankets the flames. Oil tankers, barges, storage vessels, and pipelines are quickly and safely made inert with carbon dioxide to allow for welding and other repairs.

2.4.0 Protective Measures

Material safety data sheets (MSDSs) are good sources of information on these products and must be made available to all employees of the company. There is an MSDS for each hazardous substance, listing its properties and hazards, including a hazard rating for health, fire, reactivity, and other hazards. MSDSs also detail safe handling, personal protective equipment requirements, and first aid/emergency procedures.

All products pose hazards to the environment if spilled or disposed of improperly. Areas of possible spills should have containment structures and/or procedures. These include dikes around storage and transfer facilities and containment booms around tankers being filled. The marine environment is particularly sensitive to oil spills because the tides and currents spread the oil and complex marine life is so vulnerable. *Figure 13* shows temporary containment of a spill from a tank.

Oil may be harmful on land, too. It can seep into the groundwater or streams where it contaminates plant and animal life and eventually human

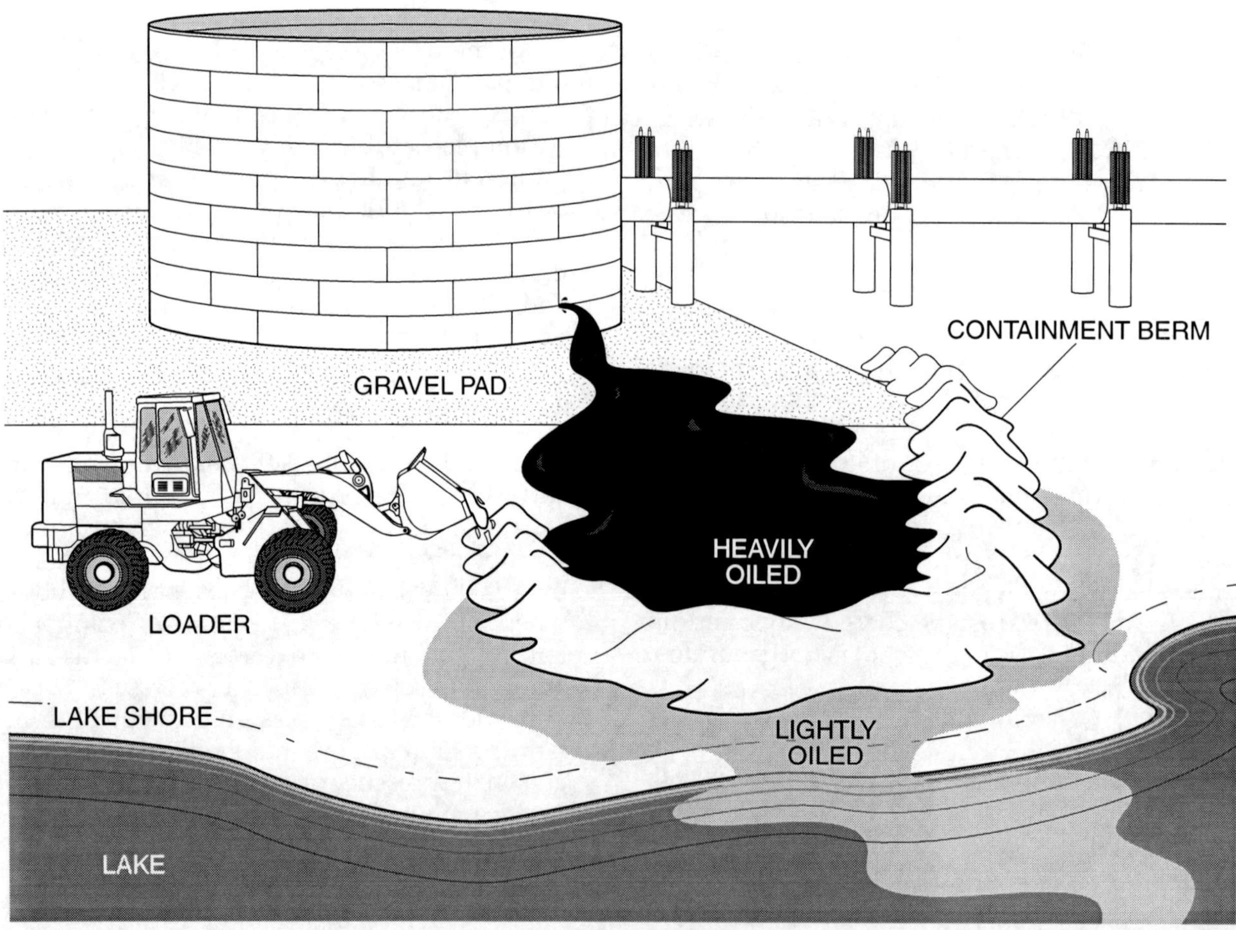

Figure 13 ◆ Temporary containment of a tank spill.

life. Handling and disposal of hazardous chemicals are addressed in company procedures and MSDSs. Detailed procedures exist for handling all petroleum products, and the observance of these procedures is strictly regulated and enforced by federal, state, and local agencies.

3.0.0 ◆ MAPS AND DRAWINGS

Maps and drawings are an important means of communicating information about pipeline equipment, flow, and processes. Some of these are piping and instrumentation diagrams (P&IDs), strip maps, blueprints, and alignment sheets.

3.1.0 Piping and Instrumentation Diagrams (P&IDs)

A P&ID depicts the functional layout of a piping system and its components, valves, and instrumentation as clearly and accurately as possible. It is accurate to the extent that all components are connected to each other as shown in relation to flow-path orientation. A P&ID does not represent the actual physical layout of equipment; for example, a valve that may appear to be right at the discharge of a pump can actually be located some distance away or even on another floor. Many times, a P&ID will use a broken line around groups of equipment to indicate that they are all in the same building (see *Figure 14*).

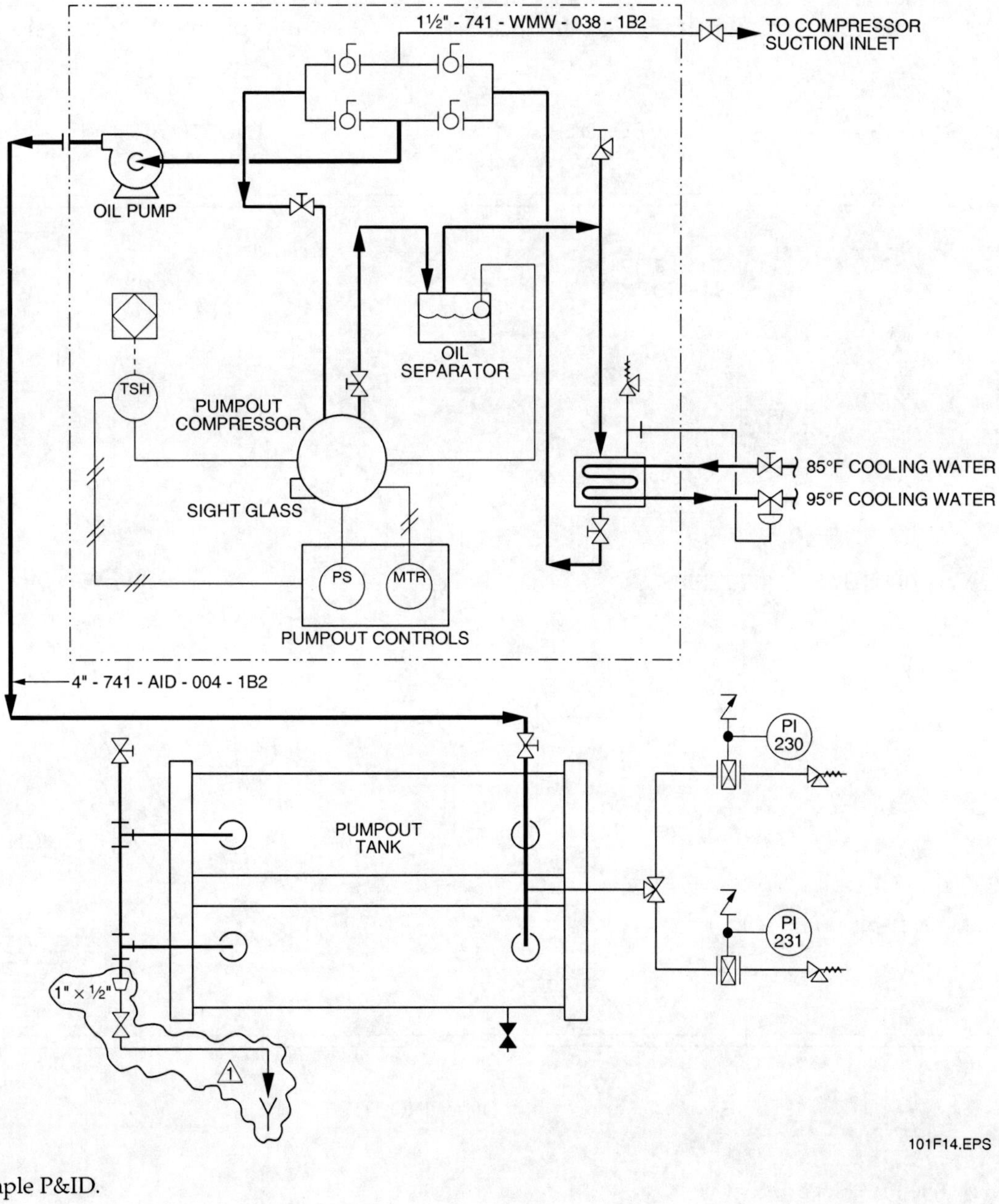

Figure 14 ◆ Sample P&ID.

The P&ID depicts all components of a particular system, including pipe sizes, flange sizes, valve sizes, flow direction, and references to other diagrams. Standardized symbols are used for these components instead of pictures. Industry standards for these symbols have been developed and adopted by organizations, such as the American National Standard Institute (ANSI), the Instrument Society of America (ISA), and the American Society for Mechanical Engineers (ASME), to aid interpretation. The P&ID title block and legend should be examined closely for explanations of the diagram.

3.2.0 Blueprints

Blueprints are duplicates, or copies, of plans for buildings, assemblies, systems, and other constructions. These are working drawings, including specifications for a project that usually includes site plans, floor plans, elevations of buildings, cross sections, and detail drawings. They were originally made by a photographic process that produced white lines against a blue background, hence, the name. However, they are now produced by computer-aided design (CAD) software and generally have black or blue lines on a white background. The blueprint shows the size and location of all structures and equipment in the project to scale and how they are physically put together. *Figure 15* shows a site-plan blueprint.

3.3.0 Strip Maps/Alignment Sheets

Strip maps (see *Figure 16*) are combinations of different types of maps or drawings over the same area such as the pipeline route. The strips of each type of map or drawing are laid out side by side

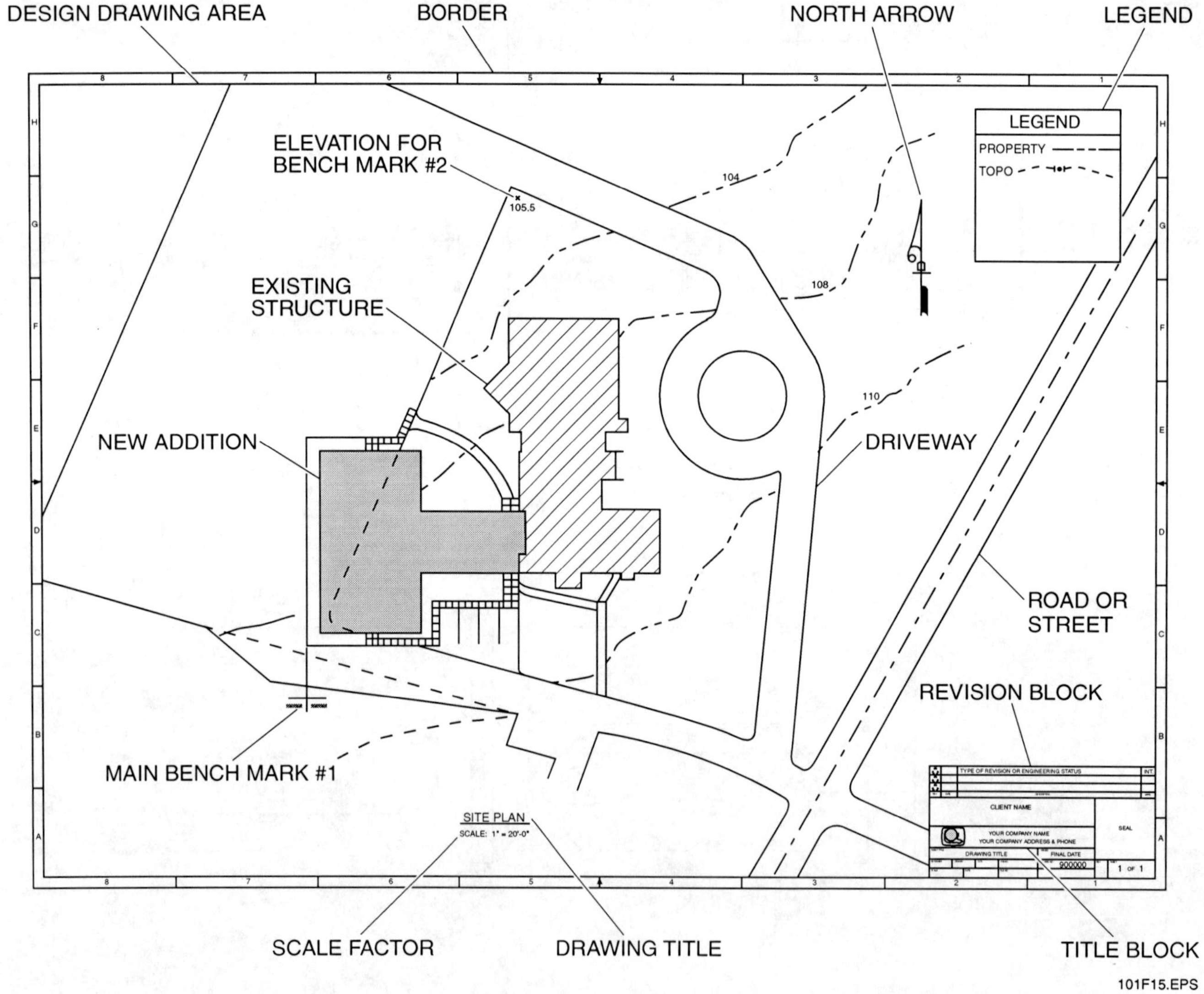

Figure 15 ◆ Site-plan blueprint.

101F16.EPS

Figure 16 ◆ Sample strip map.

on the same sheet so that they line up along the route. There may be views shown of the ground from an aerial, or plan view, and side, or elevation view, which shows the ground profile along a specific line. The pipeline will be superimposed on all views. Strip maps help to coordinate surface features with topography as well as with other pipeline data at any point on the route.

Alignment sheets are various types of maps, drawings, and diagrams that show different features along the pipeline route. The alignment sheets will show the relationship of the pipeline to true north, geographic features, political boundaries, fences, rivers, property lines, and other pipelines. *Figure 17* shows one type of alignment sheet.

Strip maps and alignment sheets vary greatly from company to company. These names may be used interchangeably or may mean very different things.

4.0.0 ◆ OPERATIONS

Controlling the fast-paced, continuous, and complex operation of the pipeline is a critical job done largely by the scheduler, controller, and dispatcher with the help of other personnel and systems like Supervisory Control and Data Acquisition (SCADA), programmable logic controllers (PLCs), and remote terminal units (RTUs). All control functions are documented by filling out forms or printing out reports.

4.1.0 Scheduling

Before a shipment of product reaches the pipeline and comes under the control of the dispatching office, it has to be accumulated and scheduled with the pipeline scheduler. The scheduler at a pipeline facility typically performs the following functions:

- Takes nominations from shippers
- Creates pumping orders for pipelines to meet shippers' needs
- Validates and reviews operating information for schedule adjustments
- Receives delivery information in time for shipments
- Validates all shipping information for billing
- Utilizes power optimization to save costs and maximize throughput efficiency

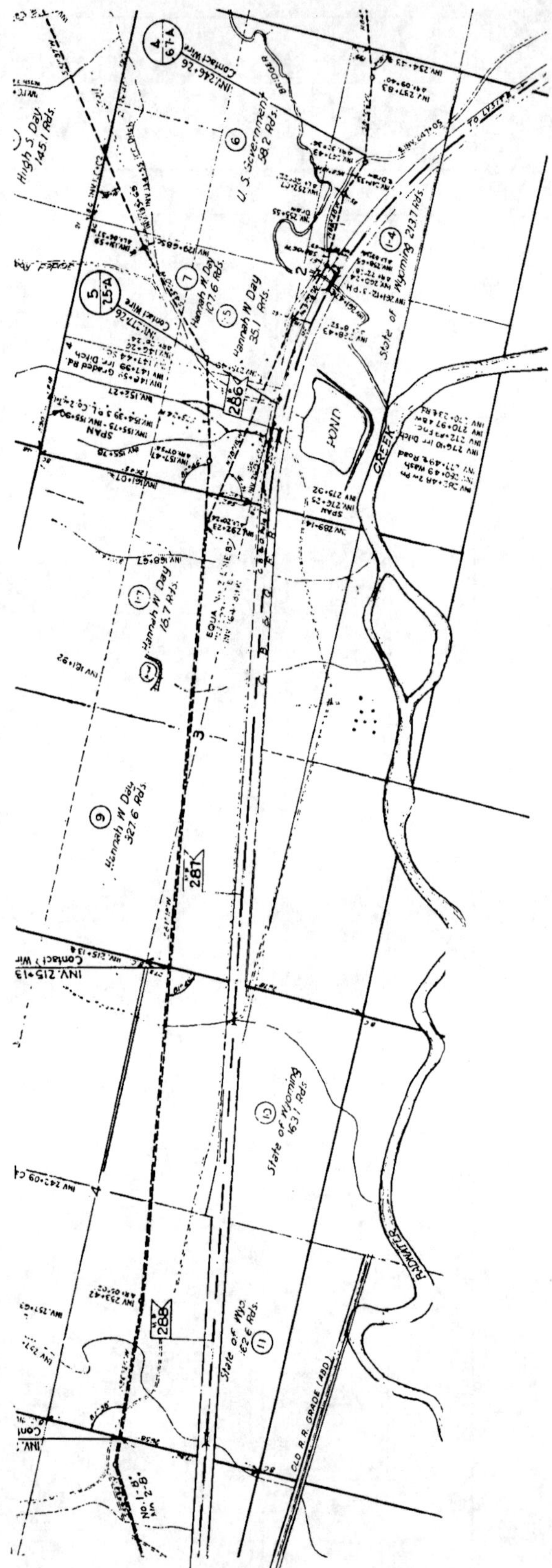

101F17.EPS

Figure 17 ◆ One type of alignment sheet.

4.2.0 Pipeline Operations

The pipeline is monitored closely and inspected for potential equipment failures and line integrity. These problems can arise from bad welds, corrosion spots, faulty valves, third party damage, or shifting of the pipe. When aboveground, the pipe may also be damaged by accidents or sabotage. In areas with faults or history of earthquakes, seismic activity must also be closely monitored. Flow can also be interrupted by obstructions in the pipe, such as stuck pigging devices, foreign objects, and debris.

Computers and automation are used to control and monitor the pipeline operation. Communication is maintained between all stations regarding the status of the operation. Controllers have the ability to shut down the pipeline flow if a problem is detected and to pinpoint the location of the problem. *Figure 18* shows a typical control center layout.

4.3.0 Control Systems

SCADA is the software used to monitor and control the flow of products through the pipeline every day. Wireless telemetry and SCADA systems provide remote alarm, control, and data collection for many types of pipeline operations.

SCADA is installed with the main computer system and hooked up through a server to RTUs and PLCs in the field. Sensors along the pipelines precisely measure operating conditions, flow rates, and the status of control devices, such as flow and pressure regulators and valves. The RTUs collect vital information in the field and transmit the data via telephone, microwave, or satellite circuits to the control center.

4.4.0 Communications

Communications are the systems that keep the pipeline operating. All along the pipeline, everyone must have open communications at all times because everything that occurs on the pipeline affects the entire system.

A variety of communications equipment and devices may be used in a pipeline company's operations. These are set up for continuity during normal or emergency conditions. It is important that communications be reliable and always available. Most companies have redundant, or backup, systems to ensure that this is true. Some of the equipment used includes:

- Dedicated satellites or microwave-based radio communications systems that provide linkage between the control room personnel and various pump stations and valve locations
- Hardwire communications that link the control room and all offices and all remotely monitored and operated facilities, such as pump stations and RTUs
- Portable cellular phones or vehicle phones provided to all key and supervisory personnel
- Locally activated pagers or nationwide pagers provided to key personnel for use in routine operating or emergency events

101F18.EPS

Figure 18 ◆ Typical control center layout.

- Fixed repeaters installed at points within the system that provide internal long distance and local radio communications

In the event of a power loss, backup systems are provided, which may include standby generators for powering communications equipment, batteries, or uninterruptible power systems (UPS). Portable self-contained mobile repeaters may be used at an incident scene and utilized to augment existing radio systems or replace any that may be inoperable. A radio link (base station) may be used with portable units to allow for emergency communications between facilities in the event that telephone and/or microwave transmissions become inoperable. Redundant communications and power systems should be installed at key facilities to allow for control in the event of loss of primary systems.

5.0.0 ◆ FLUIDS AND SYSTEM HYDRAULICS

The operation of any liquid pipeline is built on the principles of hydraulics, or the science of the behavior of fluids at rest or in motion. These principles are in turn built on the properties and characteristics of fluids.

5.1.0 Fluids

A fluid is a substance that flows, and some flow more easily than others. A fluid is also a substance that changes form with a slight force or pressure; it is elastic, however—meaning it will try to return to its previous form when the pressure is removed. Some fluids are more elastic than others. Elasticity is a measure of the fluid's resistance to change in shape.

Fluids include all liquids and gases. Water at ordinary temperature is a liquid fluid; at high temperatures, a gas fluid. Air is a gas fluid. All fluids can be greatly affected by changes in temperature and pressure.

No fluid is perfect and frictionless, and different fluids have different levels of friction, or resistance to flow. This resistance to flow is called *viscosity*.

Some fluids are thicker than others. This is called *density*. Density can be a large factor in the viscosity of a fluid. It is a measure of how tightly packed the fluid molecules are or how much of the fluid's matter exists within a given unit of volume.

Measuring density in a liquid is readily done with a hydrometer because they are not usually compressible. Measuring density in a gas is more difficult because gases can easily be compressed. Compressibility is the ability to be squeezed, or forced into a smaller volume.

Density helps determine the compressibility of a fluid. When a fluid is compressed, it becomes denser. The specific gravity of a substance is derived from its density. Specific gravity is the ratio of the density or mass of a given volume of the substance to the density of the same volume of a standard substance, such as water. The water and the liquid substance must be at the same temperature. For gases, the gas must be at the same temperature and pressure.

American Petroleum Institute (API) gravity is a unit of measure developed specifically for liquid petroleum products. The heavier the substance, the lower the number of its API gravity. Temperature will affect API gravity and should always be corrected to 60°F.

The properties of fluids that are used to describe the fluids are density, viscosity, compressibility, inertia, and elasticity. *Table 2* shows some fluid properties.

Table 2 Fluid Properties

Fluid	API Gravity @ 60°F	Specific Gravity @ 60°F	Vapor Pressure @ 100°F	Absolute Viscosity	Compressibility @ 1000 psi @ 60°F
EP Mix 80/20	214	0.41	550 psi	0.11 cP	7.3%
NGLS	146	0.51	325 psi	0.15 cP	3.8%
Raw Feed Propane	146	0.51	150 psi	0.15 cP	3.8%
Butane	112	0.58	88 psi	0.18 cP	2.0%
Anhydrous Ammonia	97	0.62	200 psi	0.25 cP	1.6%
Gasoline	63	0.66	2 psi	0.70 cP	0.8%
Jet Fuel #1	42	0.82	2 psi	1.4 cP	0.8%
Diesel Fuel #2	33	0.86	1 psi	4.7 cP	0.5%
Crude Oil	50 and below	0.78–0.98	> 1 psi	6–400 cP	> 1.0%
Water	10	1		1.12 cP	0.3%

101T02.EPS

Other properties of fluids, particularly those found in pipelines, include volatility and flammability. Volatility is the tendency of a fluid to readily vaporize. Sudden vaporization can cause a violent reaction or explosion. Flammability is a measure of how readily a fluid catches fire when in contact with heat or sparks. The volatility and flammability of specific pipeline products is described in another section of this module.

5.2.0 Basic Hydraulics

Some laws of hydraulics apply to gases as well as liquids but only under certain conditions. In general, the following sections pertain only to liquid fluids.

Hydraulics is usually divided into two branches: hydrostatics for liquids at rest and hydrodynamics for liquids in motion.

5.2.1 Hydrostatics

Hydrostatic principles are fundamental to the design of many machines and instruments. The design of hydraulic presses and the operation of instruments, such as manometers and barometers, depend on these principles.

A simple hydraulic press consists of a liquid-filled cylinder that contains two pistons, one smaller than the other. When a force is applied to the smaller piston, the force travels through the liquid to the larger piston, producing a force larger than the initial one. The increase in force is proportional to the ratio of the larger piston to the smaller one. *Figure 19* shows a simple hydraulic press.

The laws of hydrostatics are the following:

1. The amount of pressure caused by the weight of a column of liquid is determined by the height of the column.
2. An object placed in a liquid loses an amount of weight equal to the weight of the liquid it displaces.
3. A fluid in a container transmits pressure equally in all directions.

5.2.2 Hydrodynamics

Hydrodynamic principles are basic to hydraulic engineering, to the design of certain machines, and to pipeline engineering. Hydrodynamics define the plans for water systems, canals, and irrigation systems, as well as airplane design and hydraulic machines like water turbines.

Pipelines for water and other fluids have to take into account the pipe size and the fluid's friction on the pipe to determine fluid velocity through the pipe. Liquid flow can be steady or unsteady. An unsteady flow results from changes in the velocity, temperature, or pressure of a liquid or its movements around obstacles. The three basic laws of hydrodynamics apply only to steady-flowing liquids. The laws of hydrodynamics are the following:

1. The velocity of a liquid flowing through a pipe increases as the area of the pipe decreases and vice versa.
2. The pressure of a liquid increases as its velocity decreases and vice versa.
3. The velocity with which a liquid flows through an opening in a container equals the velocity of a body falling from the surface of the liquid to the opening.

Figure 20 shows liquid velocity and pressure relationships.

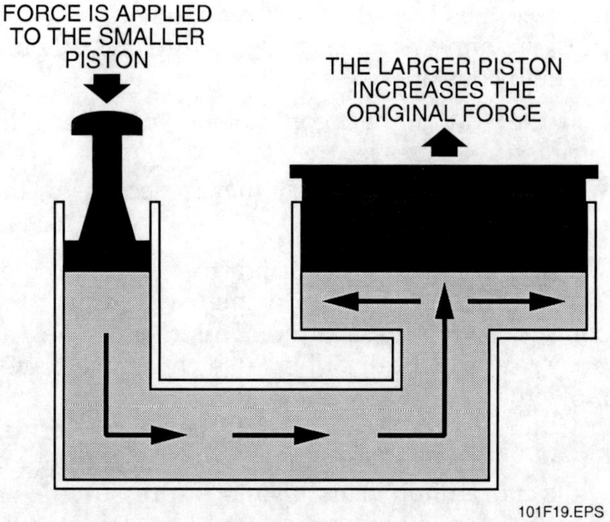

Figure 19 ◆ Simple hydraulic press.

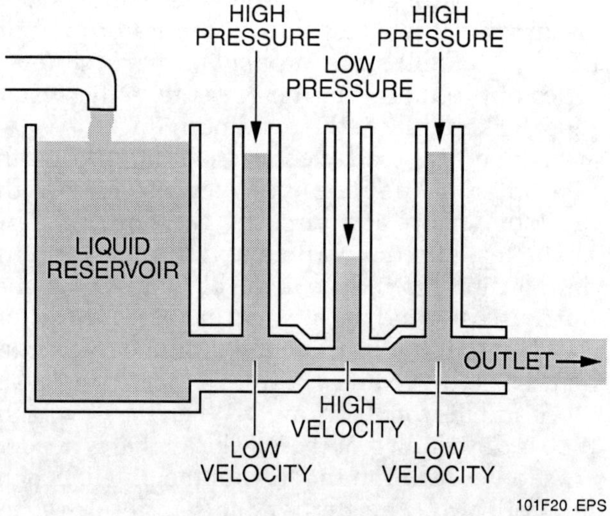

Figure 20 ◆ Liquid velocity and pressure relationships.

5.3.0 Pipeline Hydraulics

Fluid properties are major factors which affect the design of the pipeline. The primary factors in pipeline design are:

- Fluid characteristics
- Desired flow rate
- Maximum operating pressure (MOP)
- Maximum allowable operating pressure (MAOP) (for gas)
- Elevation profile/terrain
- Fittings/valves
- Installation costs
- Environmental concerns

Pressure is referenced in several ways in the pipeline industry. The maximum operating pressure (MOP) is defined as the maximum internal pressure allowed liquid during the operation of a pipeline. The term MOP is typically used more in the liquid pipeline industry and the regulations are referenced in *49 CFR 195*. The MAOP is defined as the maximum internal pressure permitted during the operation of a gas pipeline. The term MAOP is typically used more in the gas pipeline industry and the regulations can be referenced in *49 CFR 192*.

Although fluids are defined by their ability to flow, something has to start the flow and overcome their inertia and friction. A force, such as difference in pressure, gravity, or an external force physically pushing it, can do this. Usually, this is done by pumps. Stopping the flow once started will require another force, such as resistance (friction), or a stationary object, such as a closed valve.

The resistance or friction of the fluid against the pipe will be constantly trying to slow it down. Most pumps work by providing a difference in pressure and must continue providing that force all along the line to overcome the friction. There are drastic differences in the amount of pressure required to deliver the proper volume of different types of products. But the pressure in the system must be kept under the maximum limits.

Friction is also affected by the roughness of the pipe interior, the pipe diameter and length, the product density and viscosity, the flow rate, and obstacles, such as valves and bends. Pipeline design can help reduce some of these, but higher fluid viscosity and density will increase the friction.

When the fluid is traveling uphill, pumps must be used to overcome the fluid's inertia and resistance and the force of gravity. When the flow in the pipe is downhill, gravity may be all that is needed.

Another factor in the flow of the pipeline is the temperature. The higher the temperature, the lower the viscosity and vice versa. When the product is hot, it flows better. Sometimes heaters are included in the pipeline equipment to keep the product at the optimum temperature.

6.0.0 ◆ GASES AND SYSTEM PNEUMATICS

Pneumatics, as it relates to the pipeline industry, is the study of the storage, compression, and transmission of gases. The compression of gas has one primary goal, to deliver the gas at a pressure higher than the original pressure. The original pressure may vary from very low absolute pressures to several thousand pounds. The pressure rise may vary from a few ounces to many thousand pounds, and the volumes handled may vary from a few cubic feet per minute to hundreds or thousands of cubic feet.

There are three variables that have to be calculated and controlled when compressing and moving gas in a pipeline. These variables are pressure (p), volume (v), and temperature (k). These three characteristics of gases have a close relationship with one another and are explained in the ideal gas laws.

6.1.0 The Ideal Gas Laws

Pressure, volume, and temperature of gases have a close relationship with one another. By keeping one of them constant, the other two will have either a direct or inverse proportion. The relationships between pressure, volume, and temperature are explained using the ideal gas laws. These ideal gas laws are Boyle's law, Amonton's (Gay-Lussac's) law, and Charles' law.

6.1.1 Boyle's Law

Boyle's law states that when temperature is kept constant and volume is decreased, pressure will increase. His law gives the relationship between pressure and volume if temperature and amount are held constant.

- If the volume of a container is increased, the pressure decreases.
- If the volume of a container is decreased, the pressure increases.

If the volume is increased, the gas molecules have farther to travel and they will impact the container walls less often. This means the gas pressure will be less because there are fewer molecule impacts per unit of time.

If the volume is decreased, the gas molecules have a shorter distance to travel, thus striking the walls more often. This results in pressure being increased because there are more molecule impacts per unit of time.

6.1.2 Amonton's (Gay-Lussac's) Law

Amonton's law states that when volume is kept constant and temperature is increased, pressure also increases. Amonton's law was discovered by Joseph Louis Gay-Lussac in the early 1800s. His law gives the relationship between pressure and temperature when volume and amount are held constant.

- If the temperature of a container is increased, the pressure increases.
- If the temperature of a container is decreased, the pressure decreases.

If the temperature is increased, the gas molecules will move faster and they will impact the container walls more often. This means the gas pressure inside the container will increase. Gay-Lussac's law is a direct mathematical relationship. This means there are two connected values and when one goes up, the other also increases.

6.1.3 Charles' Law

Charles' law states that when pressure is kept constant and volume is increased, temperature also increases. The pressure that a gas exerts on the walls of its container is determined by the momentum of the atoms and molecules of the gas, which in turn is determined by the temperature. As the temperature increases, the atoms and molecules move faster, and so exert a greater pressure on the walls.

This law gives the relationship between volume and temperature if pressure and amount are held constant.

- If the volume of a container is increased, the temperature increases.
- If the volume of a container is decreased, the temperature decreases.

If the temperature is increased, the gas molecules will move faster and they will impact the container walls more often. This means the gas pressure inside the container will increase, but only for an instant. The greater pressure on the inside of the container walls will push them outward, thus increasing the volume. When this happens, the gas molecules will now have farther to go, thereby lowering the number of impacts and dropping the pressure back to its constant value.

6.2.0 Measuring Energy Gases

Quantities of natural gas are measured in volume units. A cubic foot of natural gas at a temperature of 60°F and an atmospheric pressure of 14.7 pounds per square inch gauge (psig) is the common unit of measure. Gas production from wells and supplies to power plants are measured in thousands or millions of cubic feet (MCF and MMCF). Resources and reserves are calculated in billions of cubic feet (Bcf) or trillions of cubic feet (Tcf). To put this in perspective, a trillion cubic feet is enough to fill a cube with sides 2 miles long.

The amount of energy that is obtained from the burning of a unit volume of natural gas is measured in British thermal units (Btus). One Btu is the amount of heat required to raise the temperature of 1 pound of water from 60° to 61°F at normal atmospheric pressure (14.7 psig). At sea level, it would take about 75 Btus to heat water enough for a cup of tea. A cubic foot of natural gas on the average gives off 1,000 Btus, but the range of values is 500 to 1,500 Btus. Therefore, 1 cubic foot of some natural gas may make only 7 cups of tea, while another would make as many as 20 cups of tea.

Energy content of natural gas varies because natural gas accumulations vary in the amount and types of energy gases they contain: the more noncombustible gases in a natural gas, the lower the Btu value. In addition, how much of any energy gas that is present in a natural gas accumulation—the mix of combustible gases—also influences the Btu value of natural gas. The more carbon atoms in a hydrocarbon gas, the higher its Btu value. To illustrate, methane typically represents more than 80% of energy gases. Methane contains one carbon atom per molecule; burning 1 cubic foot of methane gives off 1,012 Btus. Butane, possessing four carbon atoms, has a Btu value more than three times larger than that of methane. Molecular hydrogen, on the other hand, though combustible, contains no carbon atoms; its Btu value is three times smaller than that of methane.

6.3.0 Commercial Natural Gas Combustible Limits

Commercial natural gas is very flammable when sufficient amounts of oxygen (the oxygen present in air is more than adequate) are present. Mixtures of natural gas and air are explosive only when the mixture is less than 15% or more than 4% natural gas. If the natural gas is less than 4%, it will be too lean a mixture to burn, and over 15%, too rich a mixture to burn. The fact that only a small amount of gas is necessary to create an explosive mixture makes even small leaks in confined spaces very serious.

As mentioned earlier, an odorant is added to commercial natural gas so that it can be easily smelled if there should be a leak. However, it is possible for odorants to be filtered out or partially filtered out of commercial natural gas. This could happen when the natural gas passes through a filtering substance (soil or other). The presence of natural gas should be determined by testing the surrounding air with available instruments such as a combustible gas indicator or another suitable instrument.

6.4.0 Commercial Natural Gas Specific Gravity (Weight)

Specific gravity is a term used in the comparison of the weight of a gas compared to the weight of the same amount of air measured under the same temperature and pressure. Some gases may or may not be heavier than air. Air is judged to be the standard weight of one (1) with commercial natural gas weighing only .6 or 6/10 when compared to air. This means that natural gas weighs less than air so natural gas will rise up or actually float from the ground and fade away into the atmosphere when a leak occurs. Also, commercial natural gas will dissipate rapidly unless confined in a building, vault, or other structure. However, propane has a specific gravity of 1.5, is heavier than air, and will sink or fall to low places in buildings or in the ground. Butane is even heavier, weighing 2.0 or twice the weight of air, and will cling to the ground. It can be extremely dangerous when leaks occur.

Commercial natural gas is lighter than air and will normally rise and travel the path of least resistance. This means that a leak might travel uphill underground until it reaches a blockage and then travel back down the hill because that is the easiest path to follow.

7.0.0 ◆ PIPELINE EQUIPMENT

The pipeline system includes all the pipes, equipment, stations, terminals, and other facilities necessary for operation. The major pipeline equipment consists of:

- Pumps and motors
- Compressors
- Valves
- Actuators/operators
- Tanks
- Piping
- **Manifolds**
- Traps
- Measurement/sampling equipment

7.1.0 Pumps and Motors

Pumps provide the pressure and force to move product. Various types of pumps may be used in the pipeline, including centrifugal, positive displacement, hydraulic, and rotary pumps. Each is used in applications requiring different capacities or other pump characteristics. Power is supplied from many different sources, such as turbines, hydraulics, diesel engines, and electric motors. One pump station may use positive-displacement pumps driven by diesel engines, whereas another station may use centrifugal pumps driven by electric motors.

7.1.1 Centrifugal Pumps

Centrifugal pumps work by rotating and forcing product outward as product flows into the center. Revolving blades or vanes create the suction that pulls product into the pump. The product takes up the space between the vanes and rotates with them. It is then pushed outward until it leaves through the discharge, or exit, pipe. *Figure 21* shows how centrifugal pumps work.

7.1.2 Positive-Displacement Pumps

In a typical positive-displacement pump (see *Figure 22*), a piston slides back and forth in a cylinder. With each movement, it traps part of the product to be moved. This product is then moved toward the discharge end of the pump, where the pressure rises as the product is squeezed by the piston. The pressure forces the product out. There are several different types of positive-displacement pumps, including lift, or suction, pumps and force pumps.

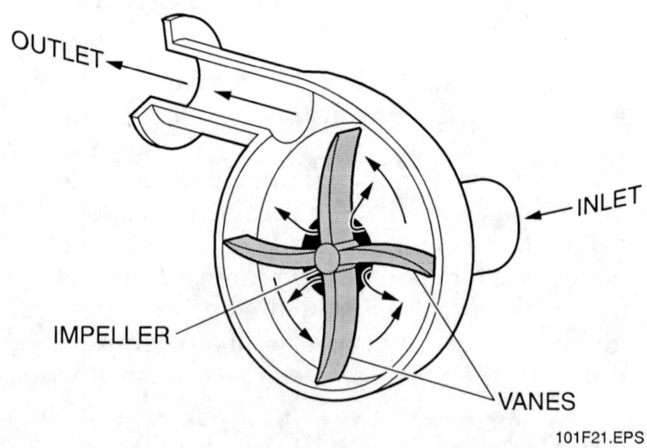

Figure 21 ◆ How the centrifugal pump works.

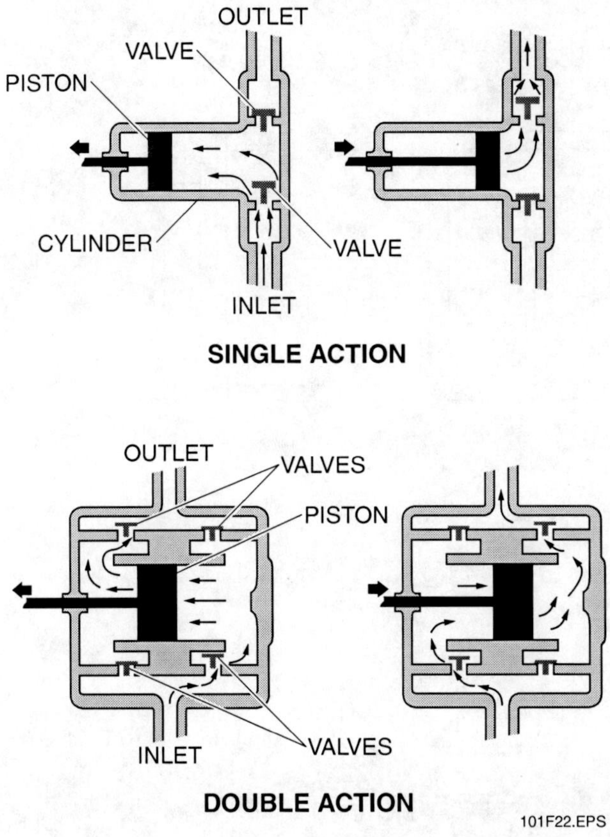

Figure 22 ◆ Two positive-displacement pumps.

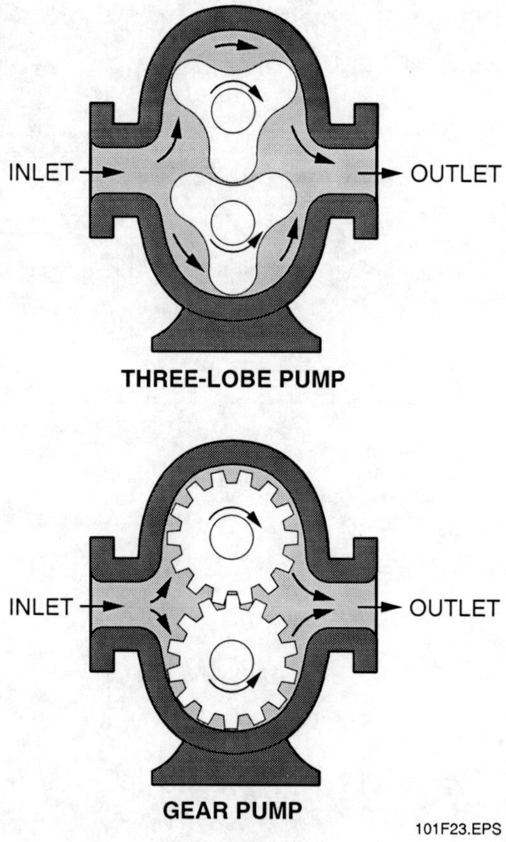

Figure 23 ◆ Two rotary pumps.

7.1.3 Rotary Pumps

Rotary pumps pump air as well as fluids. They have propeller-like blades that rotate on a single shaft. The rotation at the inlet side speeds up the fluid flow in a direction parallel to the shaft, building up pressure between the rows of blades. The pressure moves the fluid to the discharge side of the pump as the blades continue rotating.

A gear pump, which uses gear teeth instead of a piston, is used to pump thick products. The gears rotate in a housing, trapping product between the gear and the casing, moving it to the outlet side. The pressure in the outlet side builds up until it is high enough to discharge the product. Many types of drivers are used to drive these pumps. *Figure 23* shows two rotary pumps.

7.1.4 Drivers

Crude oil, refined liquid products, and gas are pushed through the pipeline grid by powerful pumps and compressors. The size of this machinery depends on the type of pipeline and the products and volumes being transported. Liquids are usually moved through the pipelines by huge pumps, while gases are moved by compressors. These compressors use pressure to push the gas through the pipeline. The pump and compressor drivers may be electric motors with fixed-speed or variable-frequency drive (VFD), diesel or gas engines, or natural-gas-fueled turbines. The choice of driver depends on the power sources available. The most common drive is a fixed-speed electric motor, although variable-frequency drives are becoming more popular. In remote locations, diesel-powered equipment is often used. Diesels are also used as backups during power failures.

7.2.0 Compressors

Compressor stations in a gas line act like pump stations in a fluid line. Compressors boost and maintain the pressure in the line, which keeps the gas flowing. In any compression process, the pressure of gas is increased from one level to another by the addition of energy to the gas being processed.

The centrifugal compressor is classified as dynamic, as opposed to a positive-displacement machine. In a reciprocating compressor, a quantity of gas is drawn into a cylinder, a piston moves, and the trapped gas is compressed by direct volume reduction or positive displacement. By this process, mechanical energy is transformed into pressure energy. *Figure 24* shows a gas compressor.

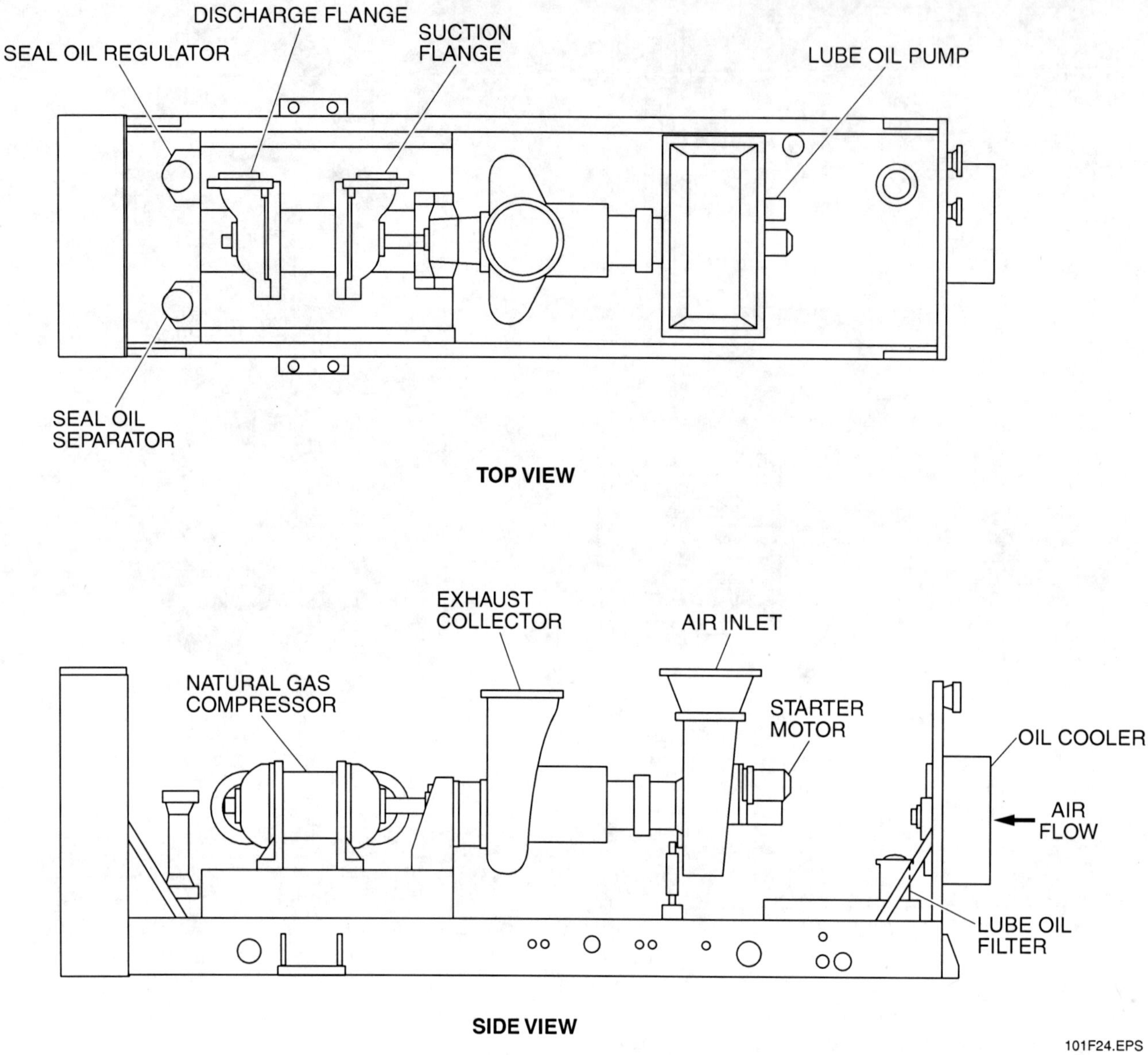

Figure 24 ◆ Gas compressor.

The combination of inlet vane, impeller, and diffuser or diaphragm is referred to as a stage. A single centrifugal stage has a relatively low pressure capability. When larger ratios are desired, additional stages are added in series. The compressor flow can be controlled and varied by changing the pressure differential across the compressor or by changing the power level. Centrifugal compressors apply inertial forces to the gas. The inertial forces are transmitted by an impeller which, by dynamic centrifugal motion, adds velocity energy through acceleration of the gas. The gas flows from the impeller into the diffuser, where the gas decelerates, and velocity energy is transformed to pressure energy.

7.3.0 Valves

Valves are a critical part of pipeline systems. They are used to control the rate of flow in a line, to open or shut off a line completely, or to serve as automatic safety devices. Valves may be identified by type or by function, and can be equipped with locking devices to prevent accidental operation of the valve. Valves may be equipped for manual operation, but most frequently operated valves in the pipeline are equipped with electric-, pneumatic-, or hydraulic-power-driven mechanical operators. Those used extensively include gate, plug, ball, butterfly, check, control, and relief valves.

7.3.1 Gate Valves

In a simple gate valve, a wedge-shaped disc or gate is pressed between inclined seats to close the valve. Operating on the same principle but with different configurations of gates and seats, there are other more complex gate valves. Conduit gate valves have an elongated gate and a port in the lower part for a fluid conduit. *Figure 25* shows a simple gate valve.

7.3.2 Plug Valves

Plug valves (see *Figure 26*) are similar in design to the stopcocks used to control flow from containers. The plug usually has an elongated port that may be smaller in cross-sectional area than the pipe. The valve is operated by giving the plug a quarter turn. Operation may be by a lever fitted to the stem, by a gear system, or by power-driven mechanical operators. Plug valves are used where reliability and fast operation are important. There is considerable sliding contact area between the plug and the valve body, and as size increases, the operating force required also increases. Many plugs have interior channels and grooves in which lubricants can be forced into the contact areas to improve turning and sealing. Plug valves are generally run in pipe in which scrapers are not run.

7.3.3 Ball Valves

Ball valves are an adaptation of a plug valve, using a ball for the plug. The sealing contact area of the ball valve is that of two seal rings. These seals are made practical by use of tough, low-friction, oil-resistant elastomers.

7.3.4 Butterfly Valves

Butterfly valves (see *Figure 27*) control product flow by use of a disc mounted on a stem that can be turned to block or allow flow. This simple lightweight valve is used in parts of the pipeline. Butterfly valves have two valuable features. They are quick-operating and take up minimal space.

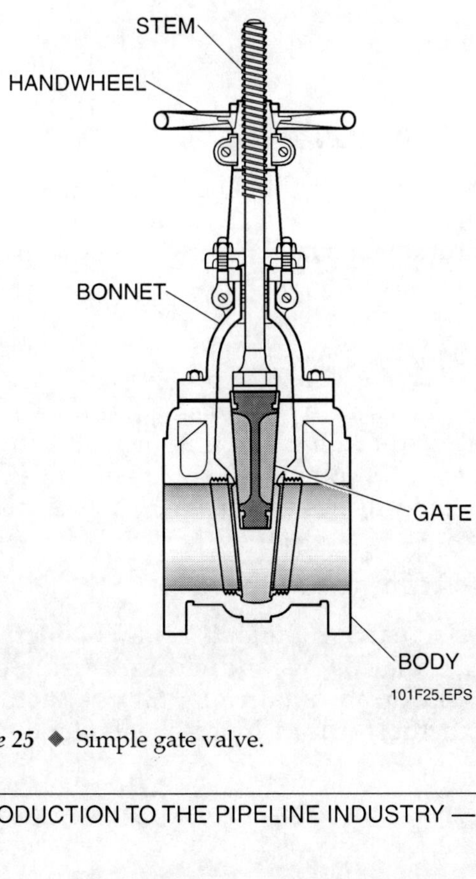

Figure 25 ◆ Simple gate valve.

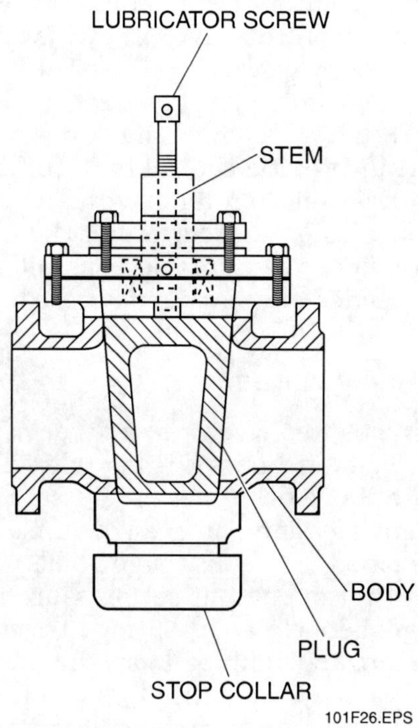

Figure 26 ◆ Plug valve.

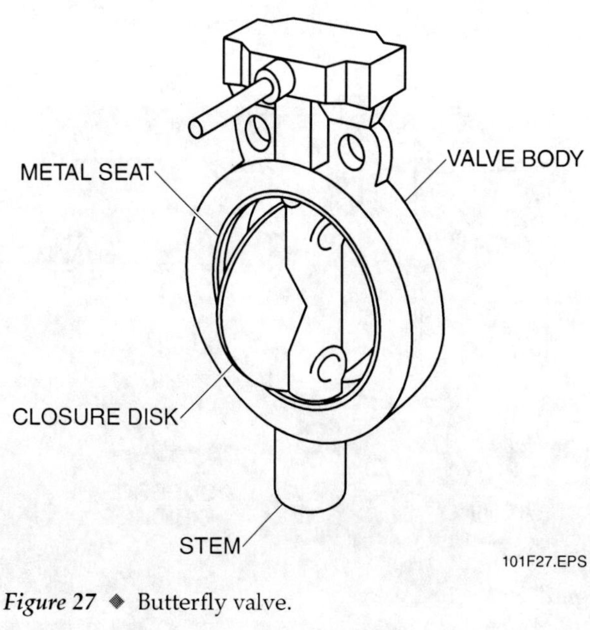

Figure 27 ◆ Butterfly valve.

7.3.5 Check Valves

Check valves (see *Figure 28*) are considered shutoff valves since they permit full flow or prevent flow in one direction. They close automatically when pressure in the flow direction is removed. Check valves can be designed to allow passage of scrapers and are most often used in gathering lines and in station piping. They are often used in conjunction with other shutoff valves.

7.3.6 Control Valves

Automatic control of flow rates and pressures is provided by control valves. They may or may not provide for complete shutoff because flow control is their major function. Port sizes and flow patterns through these valves must be continuously adjusted to provide control by throttling action, increasing their tendency toward damaging vibration, erosion, and corrosion. These valves are used principally at pump stations, although they may be needed at other points in the line.

7.3.7 Relief Valves

Relief valves (see *Figure 29*) may be called safety valves. Their function is to remain closed until pressures rise above design levels. Pressure is released to prevent pressure damage when released products are piped to safe discharge points.

When not opened at set pressure levels, the valves are kept closed by spring pressure, a nitrogen charge, a weight load, or other means. They normally close automatically when the pressure is relieved but may be designed to stay open until reset. They may be operated and monitored remotely.

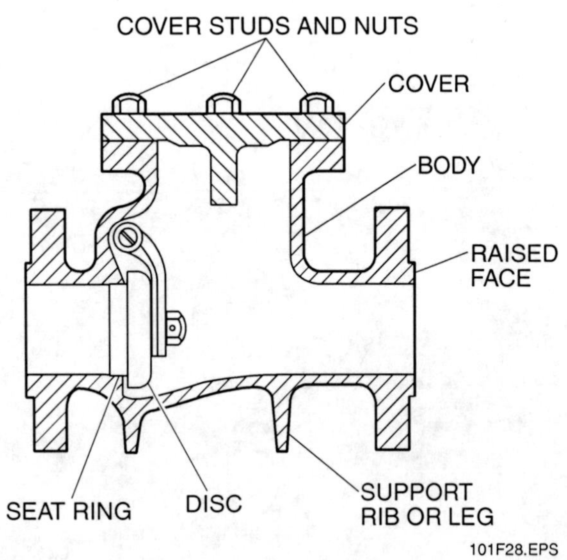

Figure 28 ◆ Check valve.

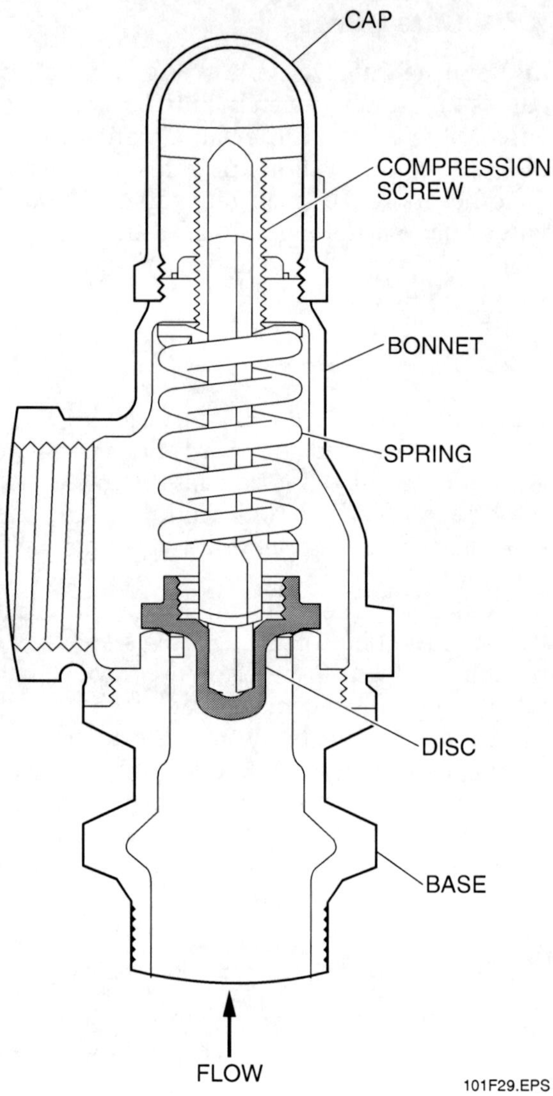

Figure 29 ◆ Relief valve.

7.3.8 Globe Valves

Other valves in the pipeline include globe valves (see *Figure 30*). Globe valves are used for shutoff and throttling, usually in auxiliary systems, such as steam and water systems.

7.3.9 Needle Valves

Needle valves are a type of small, simple, rugged device for throttling and shutting off extremely high pressures compared with other valves. They are used mostly in small auxiliary piping systems.

7.4.0 Actuators/Operators

A valve actuator, commonly called an operator, is a device that opens and closes a valve either manually or automatically. Manual actuators come in the form of handwheels, levers, and wrenches that are attached to the stem. Larger valves that require great effort to turn must be equipped with some type of powered mechanism, including electric, pneumatic, or hydraulic units, mounted on the stem. Some valve actuators use a gearbox with wide reduction ratios. Some manual actuators employ high and low gear ratios to enable personnel to turn the valve.

Valves with automatic actuators are controlled from a console in a control center by remote control, sometimes from many miles away. Such automatic actuators are used when the valve is remote from the main work area, when the frequency of operation would require unreasonable human effort, or if rapid opening and closing of the valve is required. Maintenance crews often carry portable valve actuators to work in outlying installations.

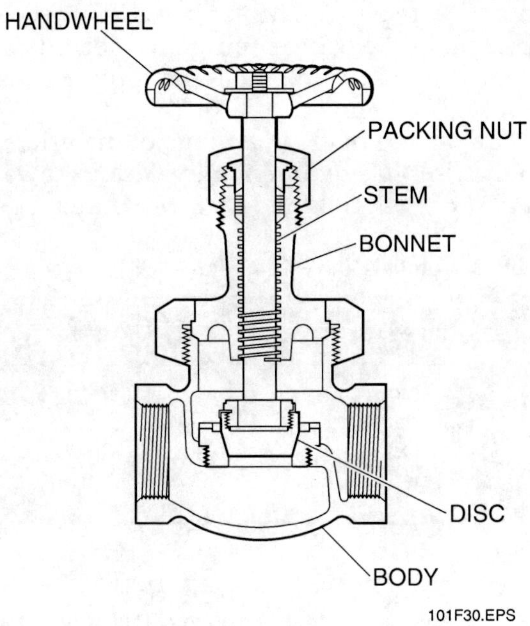

Figure 30 ◆ Globe valve.

7.5.0 Tanks

Tanks are containers used for liquid storage, relief, and breakout. Tanks may exist along the length of the pipeline, at the production site, at pump stations, at terminals, and at delivery points. Although the types and sizes of tanks at each of these sites may vary, they are usually round and made of steel. All atmospheric storage tanks are constructed per *API 650* and repaired per *API 653*.

Tanks must be calibrated for capacity. The dimensions are strapped, or measured, to be used in strapping tables. From these tables, the volume of liquid contained in the tank can be determined. Tank gauges show the tank number and gauge reading from which corresponding volume in barrels can be determined.

Cylindrical tanks may have flat floating roofs, fixed cone-shaped roofs, or both. The first can be used for gasolines and aviation-grade kerosenes, while the second can be used for fuel oils. Some tanks have fixed geodesic domes covering a floating pan. Spheroid tanks are used for volatile products because of their high vapor-pressure capacity.

The floating roof tank (see *Figure 31*) is designed to reduce losses through evaporation of volatile components of crude oil and to reduce the corrosion from sour crude by eliminating the vapor space between the oil and the roof. Seals around the perimeter of the roof prevent vapor loss and rain contamination. The wind girder on the outside rim of the tank adds strength to the shell.

Tanks must also be maintained and repaired per industry standards. Although a mixer helps to keep sediment in suspension in the product, sludge forms in the bottom and must be cleaned out periodically. The tanks are repainted inside and out at specific intervals. Structural inspections and preventive maintenance guard against internal and external damage.

7.6.0 Pressure Vessels

Just as tanks are used for the storage of liquid products, pressure vessels are used along the gas pipeline for the storage of gas products. Liquid vessels are typically cylindrical, but like tanks, can vary in size and shape from site to site, depending on application, capacity, and pressure ratings.

Cylinders and pressure vessels can be fabricated of steel, aluminum, or composite materials. They must be manufactured, inspected, marked, tested, and retested in accordance with United

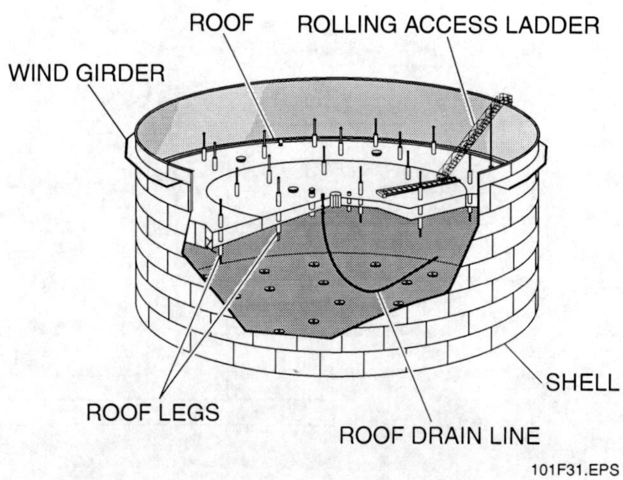

Figure 31 ◆ Open floating roof tank.

States **Department of Transportation (DOT)** regulations and exemptions for the type of service that they are used for. Pressure vessels must also be designed and repaired in compliance with *API Standards 510 and 2510*. These regulations cover the manufacture, marking, maintenance, repair, reconditioning, and testing of seamless pressure vessels, and they prescribe the requirements for transportation of hazardous materials within vessels. *Figure 32* shows a pressure vessel.

7.7.0 Pipe

Line pipe is made of steel through various methods. The main types of pipe used in the oil and gas industry are seamless, electrical resistance welded (ERW), and double submerged arc welded (DSAW). Other piping, less commonly used now, is welded steel pipe.

Seamless pipe is made from a solid length of metal pierced lengthwise by a mandrel with a rounded nose. The manufactured steel tubes and pipes are used in water, gas, and crude oil transmission and distribution lines and as casing pipes in the petroleum industry.

ERW is the most common form of manufacture for pipe in sizes from 2 to 22 inches outside diameter (OD). While seamless pipe is traditionally stronger and more expensive than comparable ERW pipe, ERW technology is improving and being used more and more frequently.

DSAW pipe is pipe that has longitudinal or spiral butt joints. The joints are produced by at least two passes, including at least one on the inside and one on the outside of the pipe. Pressure is not used, and filler metal for the inside and outside welds is obtained from the electrode or electrodes. This pipe is primarily intended for use in the gas and oil transmission industry.

Welded steel pipe is made by bending strips of steel into the form of a tube and welding the longitudinal seam either by electric resistance, by fusion welding, or by heating the tube and pressing the edges together. Two types of older piping that are still found on pipelines are furnace lap-welded and furnace butt-welded piping.

7.8.0 Fittings

Fittings include elbows, tees, reducers, flanges, plugs, and unions. Elbows are connections for directional changes. Tees are used to join three pipe ends. A reducer is used to connect pipes of different sizes. Flanges provide a means of opening and closing piping and attaching pipe ends to mating flanges on valves, tanks, pumps, and other equipment. Plugs and caps close the end of a pipe. A union provides a means of opening and closing piping and permits closing of circuitous piping systems. *Figure 33* shows types of fittings.

Fittings are made in a range of materials based on their service requirements. *Table 3* shows a partial list of materials and their pressure ranges.

Table 3 Materials and Pressures

Material	Pressure
Cast Iron	150 psig
Malleable Iron	250 psig
Cast Steel	150–2,500 psig
Forged Steel	1,000–6,000 psig

101T03.EPS

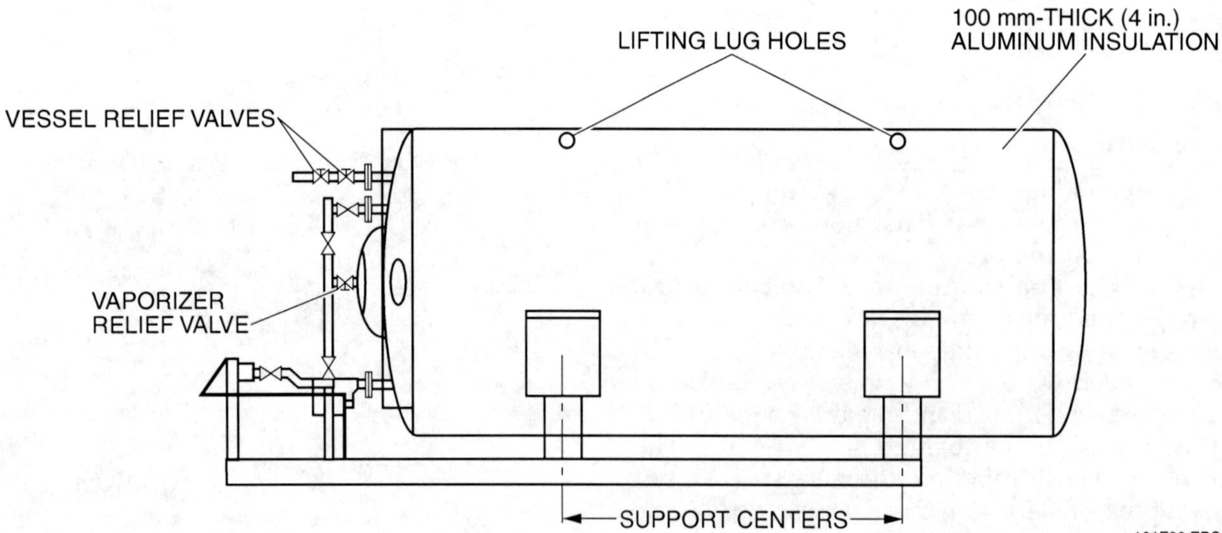

Figure 32 ◆ Pressure vessel.

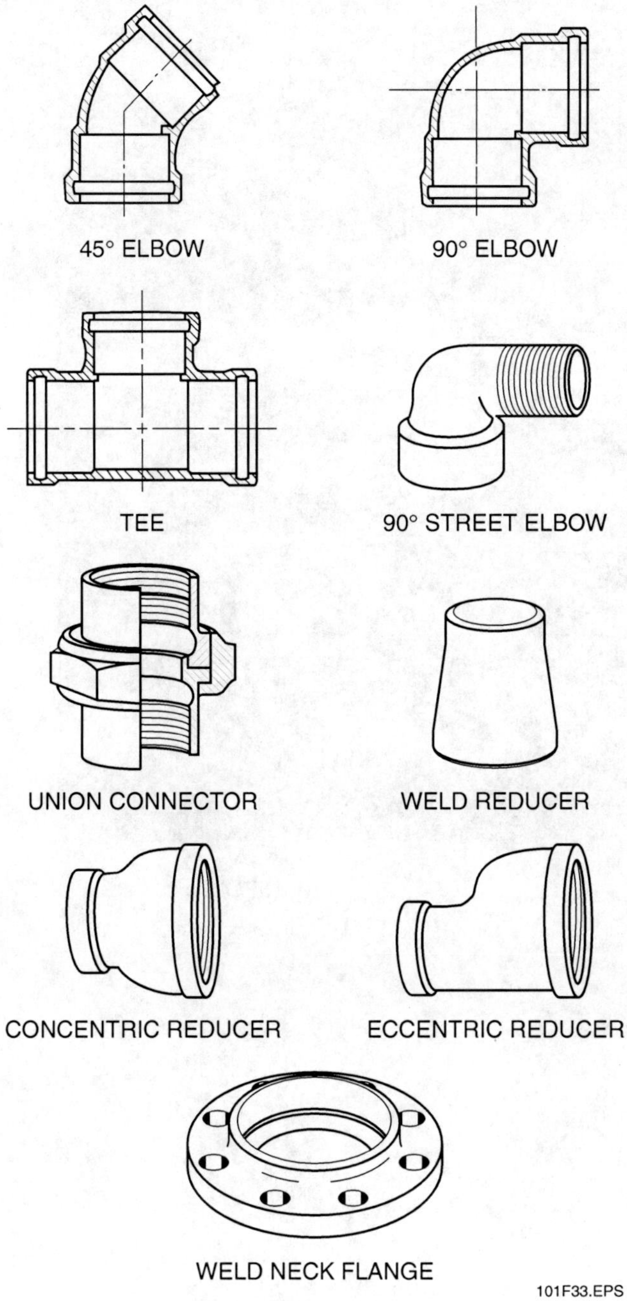

Figure 33 ◆ Types of fittings.

7.9.0 Manifolds

The pipeline uses manifolds to control the flow path of oil, gas, and other petroleum products. Manifolds, with valves and meters, branch off of the line to divert specific volumes of certain products to different locations or other pipelines. *Figure 34* shows a delivery manifold.

The complexity of manifold piping ranges from very simple to very complicated. A booster station manifold may merely allow bypassing of the station. A more extensive manifold at an originating station may connect the main line, field-gathering lines, tanks, and pumps. This manifold may perform any or all of the following operations:

- Pumping product through the main line with tanks inactive
- Receiving product from the field into any tank
- Delivering from the main line into any tank
- Transferring from one tank to another
- Pumping product from a tank into the main line
- Isolating tanks from main-line flow
- Injecting product from any tank into the main line

7.10.0 Traps

Traps in the pipeline are designed to launch or catch scrapers and other pigs that are inserted in the pipeline. Valves and/or manifolds are used to isolate the trap for scraper insertion or removal and then to realign the trap to the pipeline flow. *Figure 35* shows incoming and outgoing scraper/pig traps.

Automation of operation has been extended to the launching of scrapers in some instances. Some traps are arranged so that one or more scrapers may be loaded and then launched at any desired time by remote-controlled valve changes.

7.11.0 Measuring/Sampling Equipment

All volumes entering, leaving, and passing through the pipeline system must be measured and accounted for. Accurate measurement is required for custody transfer, task purposes, leak detection, and delivery schedules.

The primary method of measurement in the pipeline is metering. Meters, whose accuracy is documented with meter provers, measure and record the volume of product passing through. Factors affecting volume are the temperature, pressure, and density of the product being measured. Various instruments are used to measure these factors concurrently with the volume measurement. Product characteristics are also checked periodically with automatic and manual sampling.

7.11.1 Meters

Various meters are used along the pipeline to monitor the volume of fluid or gas. Most product measurements are made by turbine meters or positive-displacement meters, although coriolis-type mass flowmeters, vortex, and ultrasonic meters are also used. Meter readings taken before

Figure 34 ◆ Delivery manifold.

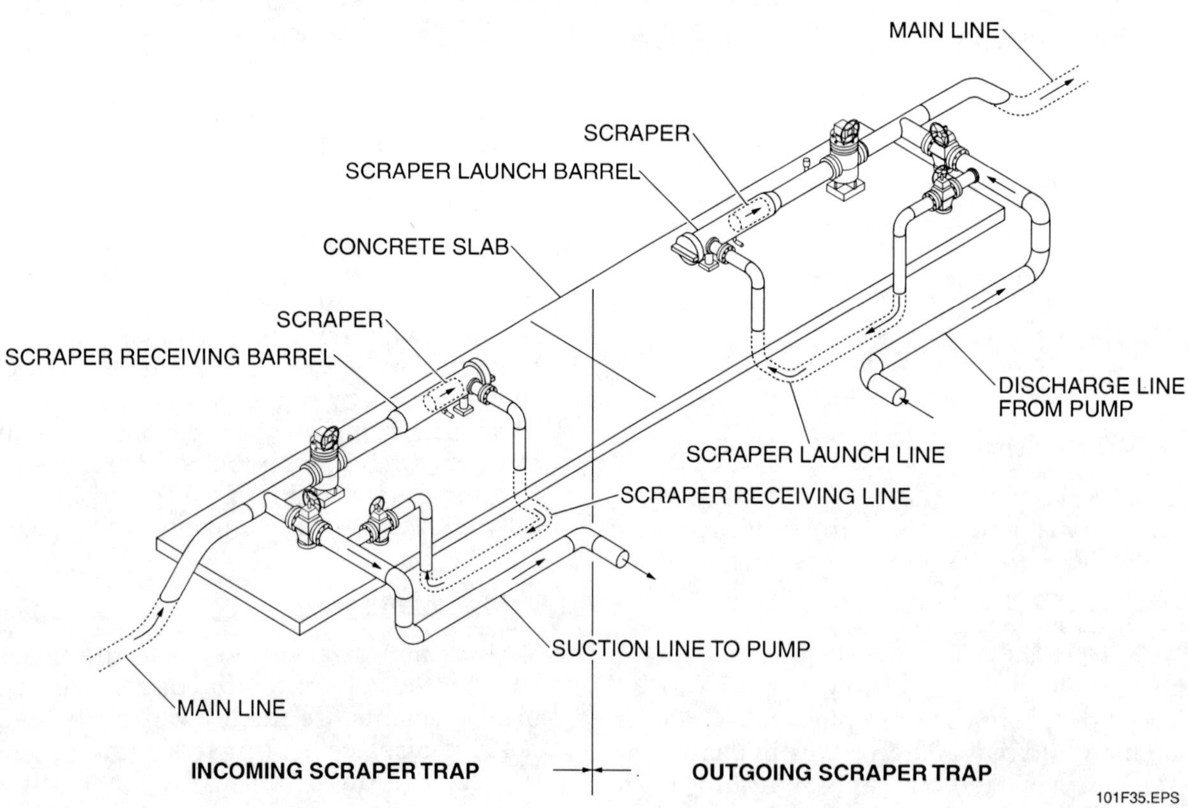

Figure 35 ◆ Incoming and outgoing scraper, or pig, traps.

and after deliveries are used to determine the volume delivered, or **throughput**. Meters are typically equipped with transmitting, remote-reading, and print-out accessories to facilitate the delivery of correct product volumes and for preparation of delivery tickets. Data is often transmitted to remote locations. *Figure 36* shows two types of meters.

Gas is typically measured by orifice, turbine, positive-displacement, ultrasonic, and rotary vane meters. These meters are generally automated to correct gas volumes for moisture content and temperature and to provide remote read-out and printing capability. Strainers are installed upstream of meters to remove any foreign material from the stream before it goes through the meter. The wire mesh in strainers is generally more coarse than filter materials are.

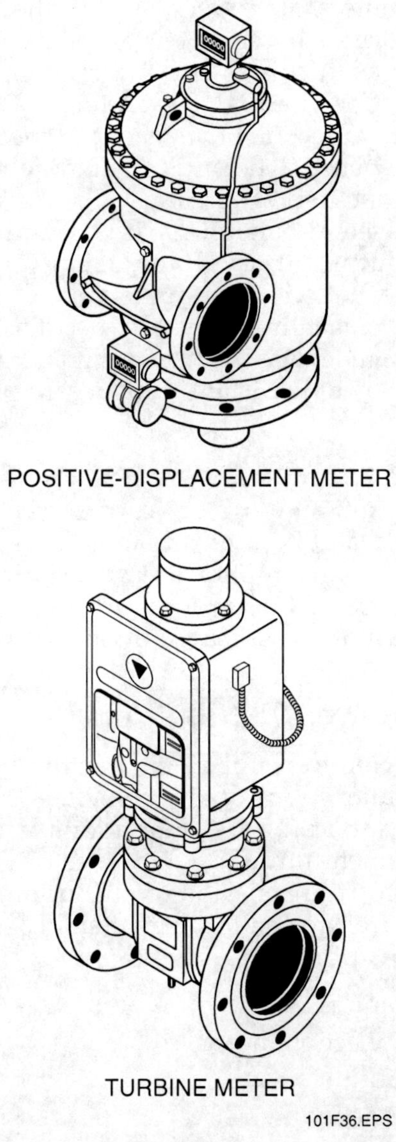

Figure 36 ◆ Two types of meters.

The orifice plate is the vital component of the orifice meter. It is the part that restricts the flow to develop the difference in pressure that is measured to calculate actual flow. The accuracy of orifice meters is dependent on the quality of the meter that is installed. Expert installation and maintenance are necessary to ensure the continued accuracy of the meter.

Orifice plates are secured between halves of flanges of machined carriers or in the body of a special orifice meter housing constructed so that the plate can be removed for inspection. Some meters allow the plate to be removed without stopping the flow. The orifice meter is made up of the carrier housing or flanges, the orifice plate, gaskets, O-rings, devices for removal, and pressure fittings. Pressure taps are drilled at a location exactly 1 inch on center from either side of the orifice plate. *Figure 37* shows some common orifice plate holders designed for easy removal.

Visual inspection of the housing unit involves close examination of the unit looking for any signs of rust, corrosion, cracks, or any buildup of foreign material around the outside housing joints, suggesting the possibility of a leak. Pressure fittings and sensor connections must also be inspected for leaks or damage of any kind. Plates vary from 2.5 inches to 30 inches in diameter.

7.11.2 Provers

A meter prover is used to calibrate meters for specific conditions of flow and product type. Meters are proved by passing a volume of product through the meter at a known rate and comparing this metered volume with the known volume prover. Two types of provers frequently used are the ball prover and the piston prover. A prover loop is a length of pipe or a calibrated U-shaped pipe with a known volume used to check or prove the accuracy of a meter.

8.0.0 ◆ ELECTRICAL POWER SYSTEMS

The electrical power system used by any pipeline company may be quite large or quite small. Power may be self-generated, or partially so, or may be purchased from another source. For the purposes of an introduction, a basic power system will be described from generation to delivery.

An electrical power system is an assembly of equipment and circuits for generating, transmitting, and distributing electrical energy. The elements of a power system form complex networks that require energy control centers to monitor and regulate their operation. The reliability of the system is critical to pipeline operation. *Figure 38* shows an electrical power system. The principle elements of the system can be divided into the following:

- Power generation and transmission
- Power distribution
- Direct current (DC) power distribution
- Utilization equipment

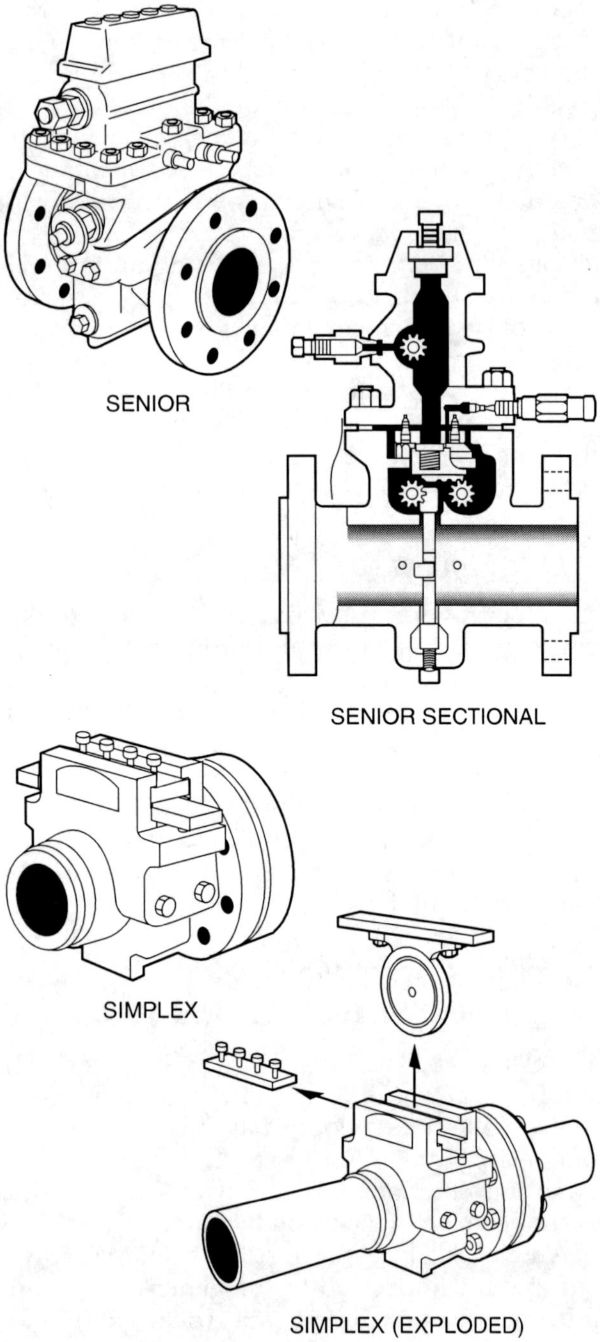

Figure 37 ◆ Common orifice plate holders.

8.1.0 Power Generation and Transmission

Electrical systems require generating stations to produce electric power, and transmission systems to carry it to areas of consumption. Along the way, the voltage of the power delivered must be adapted to the specific needs of consumers. The myriad of uses that consume this energy is the power system load.

8.1.1 Generation

The large quantities of electricity required to supply a system are produced in generating stations, commonly called power plants. These are actually conversion facilities where the chemical or nuclear energy of fuel, such as coal, oil, gas, and uranium, or the kinetic energy of falling water is converted to electricity.

The generators, which supply the electricity, can be driven by steam turbines, hydraulic turbines, gas turbines, or internal combustion engines. Coal, natural gas, and heavy fuel oil are commonly used fuel for steam turbines.

Three-phase 60 hertz (Hz) AC systems are used almost exclusively in the U.S. Consequently, power system generators are wound for three-phase output at a voltage usually limited by design features. The output of modern generating stations is usually stepped up by transformers to the voltage level of transmission circuits used to deliver power to distant load areas.

8.1.2 Transmission

The transmission system carries electric power efficiently and in large amounts from generating stations to consumption areas. Such transmission is also used to interconnect adjacent power systems for mutual assistance in case of emergency.

8.2.0 Power Distribution

Power is reduced and distributed through a series of substations to provide the voltage required in specific applications. A substation is an assembly of equipment through which electrical energy is passed for transmission, distribution, interconnection, transformation, conversion, or switching. Power distribution involves:

- Substations
- Disconnection means
- Transformers
- Cables/conduits
- Switchgear
- Motor control centers

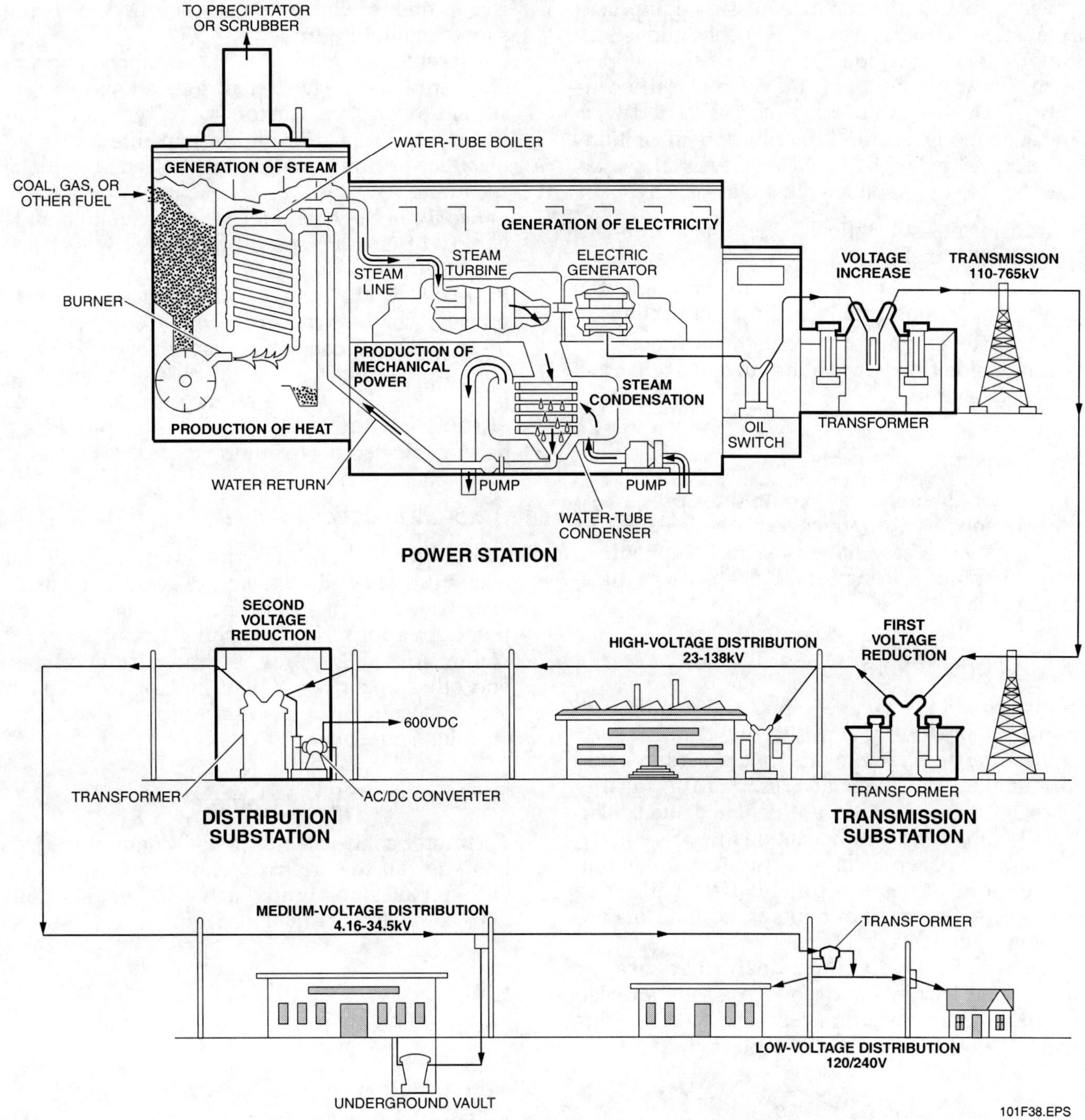

Figure 38 ◆ Electrical power system.

8.2.1 Substations

Substations are used for some or all of the following purposes:

- Connecting generators, transmission or distribution lines, and loads to each other
- Transforming power from one voltage level to another
- Interconnecting alternate sources of power
- Switching for alternate connections and isolating failed or overloaded lines and equipment
- Controlling system voltage and power flow
- Compensating for reactive power
- Suppressing overvoltage
- Detecting faults, monitoring, and recording information, power measurements, and remote communications

Minor distribution or transmission equipment installation is not referred to as a substation. Substations are referred to by the main duty they perform. Broadly speaking, they are classified as transmission substations when associated with high-voltage levels, and distribution substations when associated with low-voltage levels. However, they are also referred to in other ways, such as:

- Transformer substations, when the equipment includes transformers
- Switching substations, when the equipment is mainly for connections and interconnections
- Converter stations, when the main function is conversion from alternating current (AC) to DC and vice versa

8.2.2 Disconnection Means

Disconnect switches are commonly present so that sections or components can be isolated for maintenance. These have isolation and connection capability but may not be capable of interrupting load.

8.2.3 Transformers

Transformers use magnetic core and windings to transform power from one side of the transformer to the other at different voltages. Substation transformers range from small sizes of 1MVA to large sizes of 2,000MVA. Most are insulated and cooled by oil, and adequate precautions must be taken against fire hazard. These include adequate distance from other equipment, fire walls, fire extinguishing means, and pits and drains for containing leaked oil.

Current and potential transformers are for measuring currents and voltages. They are also for providing proportionately low-level currents and voltages for control and protection.

8.2.4 Cables and Conduits

Cables are two or more insulated wires and a grounding wire covered by an outer jacket, or sheath. Cable is usually classified by the type of covering it has, either nonmetallic or metallic. Cable may also be classified according to where it can be used. The *National Electrical Code* has definite guidelines on cables and their allowable uses. Because water can cause power loss or short circuits, cables are classified for either dry, damp, or wet locations. They are also classified for exposure to sunlight or rough use.

All cables are marked to show important properties and uses. These markings show the wire size, number of conductors, cable type, and voltage rating. The information is printed on nonmetallic sheathing and on tags attached to metal sheathing.

Raceway is a general term that refers to a wide variety of circularly and rectangularly enclosed channels used to house electrical wiring. Conduit is a raceway with a circular cross section similar to pipe that contains cables or wires. Conduit is used to provide protection for conductors and route them from one place to another. In addition, conduit makes it easier to replace or add wires to existing structures and, if it is metal, provides a permanent electrical path to ground.

8.2.5 Switchgear

Switchgear includes a wide variety of enclosed circuit breakers, disconnects, busways, and associated items that distribute power from a common point. An appropriate switching arrangement for connections of generators, transformers, lines, and other major equipment is basic to any substation design. Switchgears are use to provide this arrangement.

8.2.6 Motor Control Centers

The motor control center (MCC) contains the isolators for all the electrical equipment associated with a main line pump. Each MCC **cubicle** contains some or all of the following:

- Breaker
- Insulator
- Disconnect
- Isolator
- Reset button
- Hand/auto switch
- Run and stop light indicators
- Opened, closed, and stop light indicators

The isolator is the main isolator for the identified equipment and may be padlocked in the OFF position. The run and stop light indicators display the status of the associated equipment. Opened, closed, and stop light indicators are associated with motor-operated isolation valves.

8.3.0 DC Power Distribution

The DC power system provides the controls, monitor, and indicators of power required for control systems. Normal shutdown and automatic malfunction shutdown functions are also provided in the electrical system.

The DC power supply consists of batteries and chargers. A source of DC power is externally supplied by batteries and a battery charger. The charger and certain other equipment require an external power source. The battery charger will automatically charge the batteries. It does so by changing the batteries to full charge and then provides a float charge that will preserve the charge.

The charger will carry continuous and intermittent DC loads up to 80% of the maximum rated output. The charger has limiting characteristics to avoid exceeding its maximum output rating, and the output is further protected by fuses for short-circuit conditions.

9.0.0 ◆ CORROSION CONTROL

Corrosion is a natural process that, given the right conditions, can attack any metal or alloy. Most pipeline technicians are familiar with corrosion but do not have specific knowledge of the various types of corrosion. Everyone is surely familiar with rust, the most common form of corrosion which results in overall thinning of a metal. Most corrosion failures, however, are not caused by rust or general corrosion but by localized attack—pitting, crevice corrosion, intergranular corrosion, or stress corrosion cracking.

Each year unchecked corrosion can cost the pipeline industry millions of dollars in material damage and downtime. Corrosion can cause catastrophic failures with potentially severe consequences for the environment and serious injury or even death to personnel. It is necessary today to be able to identify potential problems and risks through a corrosion management program. An entire industry has built up around this need.

The largest organization in the world committed to the study of corrosion is the National Association of Corrosion Engineers (NACE). NACE was established in 1943 by eleven corrosion engineers in the pipeline industry. Today, NACE International has more than 50 years of experience in developing corrosion prevention and control standards. NACE is an excellent source of information for further study.

Federal guidelines have been implemented to support safety of both personnel and the environment. The complete text of the corrosion control requirements can be found in *49 CFR Part 192, Subpart I* and *49 CFR Part 195, Subpart H*. Companies are required to have written guidelines and procedures for most corrosion-related activities.

9.1.0 Types and Causes of Corrosion

Many factors contribute to metal corrosion; dissimilar metals that corrode at different rates and are sometimes incompatible, dissimilar soils, dissimilar aeration, bacteria, and stray current are factors that contribute to corrosion.

Pipeline corrosion is most prevalent when the failure of coatings, inhibitors, or cathodic protection occurs in a corrosive environment. These factors do not affect the pipeline equally at all locations, and corrosion does not grow at the same rate throughout a pipeline.

Internal corrosion of oil and gas pipelines is a growing concern to pipeline operators. The causes of internal corrosion of pipelines are chloride, carbon dioxide, hydrogen sulfide, oxygen, and microbiological activity. A recent severe case of internal corrosion caused several failures on a 20-inch oil line after only 2 years of service. The cause of the failure was high carbon dioxide concentrations (9%) in combination with high levels of brine.

Piping and pipeline systems are often great places to grow microbes that can produce corrosive conditions. Microbiologically influenced corrosion (MIC) often works in combination with other forms of corrosion, such as crevice corrosion and under-deposit corrosion, on in the pipe. It can also accelerate corrosion produced by impurities in the system environment, such as hydrogen sulfide and/or carbon dioxide.

For example, MIC can occur in the crevices produced by disbonded coatings on the outside of the pipe that are shielded from cathodic protection. Internally, sites for MIC are where water, debris, and other nutrients accumulate on the pipe wall. Both sites provide the local conditions that promote colonization of microbes, which can make existing corrosion problems worse by producing organic acids or acid-producing gases. There is a long list of microbes that can participate in the corrosion reaction; these include aerobic or anaerobic types (depending on whether they require oxygen or lack of oxygen to survive). They can also make use of nutrients in the environment (sulfates, nitrates, iron) to produce corrodants.

9.2.0 Inspection and Testing

Pipeline integrity programs use tools, technologies, and strategies to ensure that pipeline systems are in a condition to perform safely and reliably. Increasingly sophisticated technology is available. Constant monitoring and inspection is essential to any pipeline integrity program.

9.2.1 Microbiologically Influenced Corrosion (MIC)

A variety of methods is available to detect MIC. Sampling and culturing can be performed for identification of the size of the infection and the active strains of organism present. This information is then used to identify preventative or remedial treatments. Chemical analysis and pH of corrosion products can also give important information relative to the nature of the problem. Evidence taken from the metal surface, called sessile surveys, has a better chance of properly assessing the situation versus sampling of the liquids, called planktonic survey.

9.2.2 Sampling

Injection and sampling systems are fundamental to corrosion control and process control programs. They are applicable to a large variety of processes in the petroleum, chemical, and water treatment industries. Injection systems are used to inject a wide range of chemicals into processes. Such chemicals include biocides, demulsifiers, corrosion inhibitors, oxygen scavengers, glycol and monoethylene glycol, dewaxers, methanol, odorizers, and product additives.

Sampling systems are used to take samples of the process product. Such samples are then analyzed in the laboratory for inhibitor concentration levels, the presence of metal ions, oxygen levels, scale forming compounds, and a wide range of other substances.

9.2.3 Corrosion Coupons

Coupons are specimens of materials that are exposed to tests or real environments to assess the effect of degradation on the material. They can be used inside a pipeline or outside if the pipe is underground. Corrosion coupons represent a widely used and relatively simple form of corrosion monitoring. Usually a coupon is exposed to a corrosive environment to represent an actual component/structure. After exposure for a specified amount of time, the coupons can be weighed and examined microscopically to determine and evaluate the type and average rate of corrosion damage. *Figure 39* shows typical coupons.

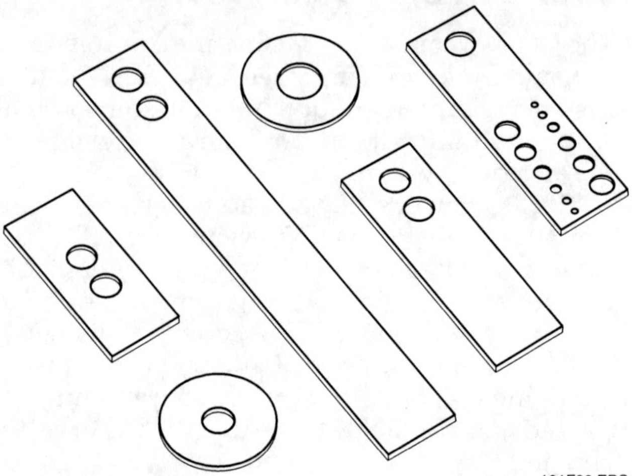

101F39.EPS

Figure 39 ◆ Typical coupons.

9.2.4 In-Line Inspection, or Smart, Pigs

Smart pigs are high-tech inspection tools covered with sensors that gather and record information in onboard computers as they pass through the pipeline. They check for various problems, such as thinning, pitting, and general corrosion. Some smart pigs can detect metal loss in areas around welds. *Figure 40* shows an ultrascan in-line inspection (ILI) pig.

9.3.0 Maintenance and Protection

Monitoring a pipeline for corrosion is essential in preventing failures. If an operator can identify active or growing corrosion defects, then the rate of corrosion can be individually predicted for each defect, allowing an opportunity for corrective measures before the failure occurs.

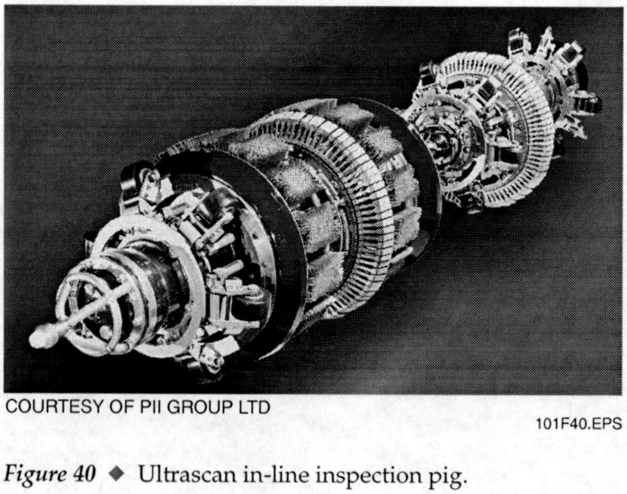

COURTESY OF PII GROUP LTD

101F40.EPS

Figure 40 ◆ Ultrascan in-line inspection pig.

As soon as production begins in a line, companies begin maintenance programs to reduce the risk of pipeline leaks. Sophisticated electronic control systems monitor changes in pressure, volume, and product flow. A variety of highly technical tools collect information about both the interior and exterior condition of pipelines. In-line inspection (ILI) tools, commonly called smart pigs, assess metal loss, strain, and cracking. This information becomes part of a computer database that is used by experts to identify and plan long-term maintenance programs.

To safeguard the outside surface of the pipe, protective coatings are applied when lines are built. In all cases, as a backup to the protective coating, cathodic protection is installed to help control external corrosion.

9.3.1 Inhibitors

A corrosion inhibitor is a chemical substance that, when added in small concentration to an environment, effectively decreases the corrosion rate. Inhibitors are chemicals that react with a metallic surface or with the environment the surface is exposed to, giving the surface a certain level of protection. Inhibitors often work by adsorbing themselves on the metallic surface, protecting the metallic surface by forming a film. Inhibitors are normally distributed in a solution and injected into a pipeline through a sampling/injection tool. Some are included in a protective coating formulation. Inhibitors slow corrosion processes by one of the following methods:

- Increasing the anodic or cathodic polarization behavior
- Reducing the movement or diffusion of ions to the metallic surface
- Increasing the electrical resistance of the metallic surface

A vapor corrosion inhibitor stops corrosion by releasing protective ions that are attracted to metal. These ions condense on the metal surface, forming a water-repellent layer that contains both positively and negatively charged ions. The layer is maintained and repaired by the constant condensation of new vapor.

Corrosion inhibitors used in deep-water treating of oil and gas wells and deep-water risers differ from the use of land-based inhibitors in several ways, including delivery of the chemical to the point of injection, selection of inhibitor, and monitoring for results.

Each situation must be carefully examined before the proper inhibitor is selected.

9.3.2 Cathodic Protection

Cathodic protection effectively protects underground or submerged metallic structures through the use of a negative potential applied by an external source to the pipeline or structure. Usually, once the pipeline has been made sufficiently negative, it will resist environmental (soil or moisture) corrosion. The method is typically applied to underground pipelines, storage tanks, the interior of water storage tanks, ocean pilings, and electrical transmission towers. Buried steel structures will revert back to their natural state as an iron oxide without proper protection.

Cathodic protection is the most widely applied electrochemical corrosion control technique that reduces the corrosion of a metal surface by making that surface the cathode of an electrochemical cell. Cathodic protection is accomplished by creating and controlling the action of large corrosion cells by connecting the negative terminals of DC power sources to the pipeline and the positive terminals to groundbed anodes (see *Figure 41*).

10.0.0 ◆ REGULATIONS, PROCEDURES, AND DOCUMENTATION

Each pipeline company has published operations and maintenance (O&M) and emergency procedures. These procedures are based on federal, state, and local regulations and on industry standards. Regulations have the effect of laws and are enforced by government agencies. Standards are generally accepted practices gained through experience. All pipeline employees are expected to become familiar with all procedures that are relevant to their job.

10.1.0 Regulations

The U.S. Department of Transportation (DOT) Office of Pipeline Safety also handles pipeline safety and the quality of pipeline design, construction, and operation. This office administers two sets of regulations, *Code of Federal Regulations (CFR) 49 Part 192* and *CFR 49 Part 195*, that address transportation of natural gas and other gases and liquids by pipeline. The pipeline industry is greatly affected by federal regulations. The Federal Energy Regulatory Commission (FERC) controls the natural gas and crude oil and products lines in addition to the regulations of other federal agencies. The Occupational Safety and Health Administration (OSHA) oversees every facet of worker safety and health, from the wearing of personal protective equipment to provision of hazard information to availability of emergency equipment.

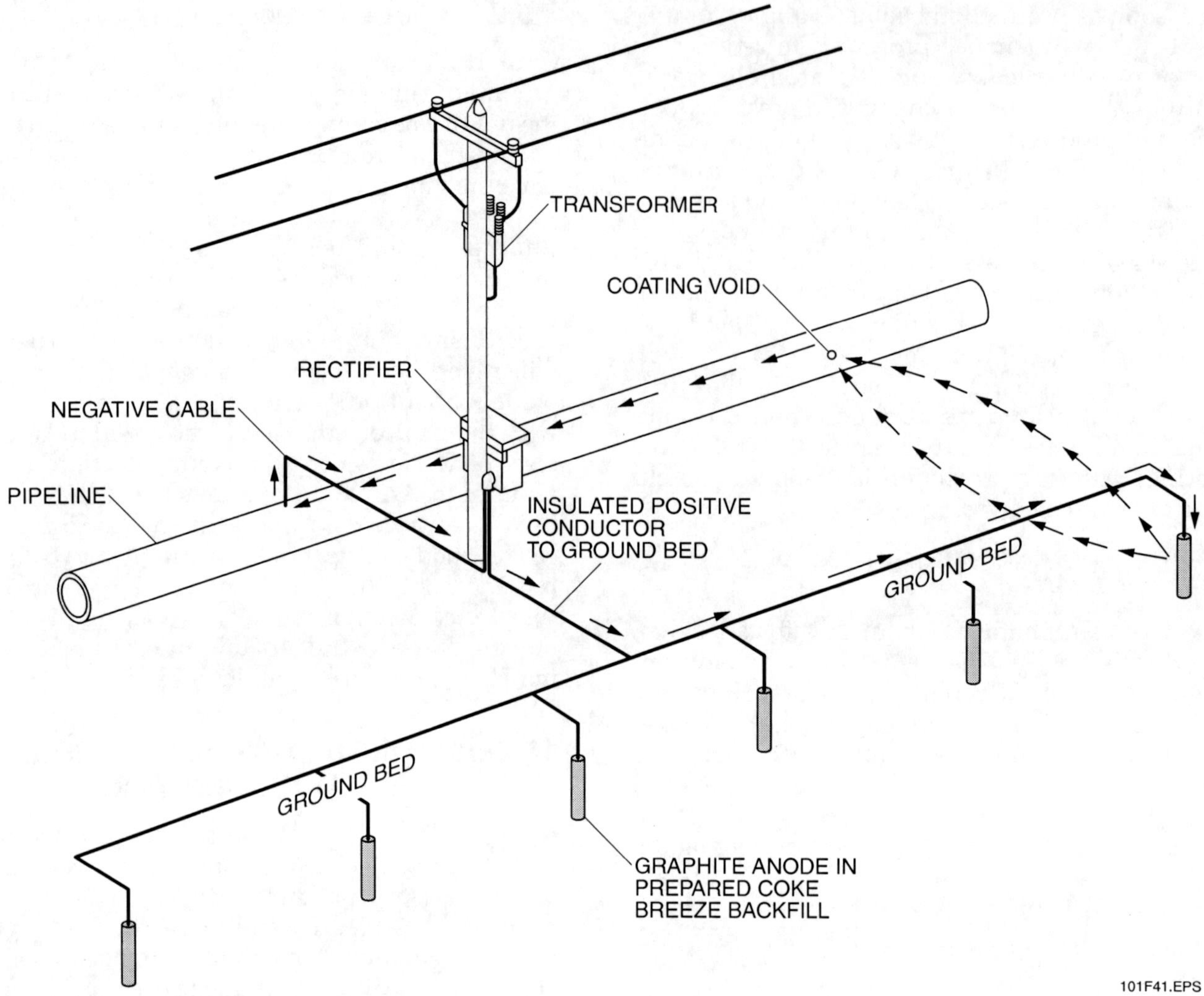

Figure 41 ◆ Cathodic protection arrangement.

State agencies can adopt federal regulations or enforce more stringent regulations than federal because each state has developed their own form of regulation regarding pipelines. Pipelines operate on an interstate or intrastate basis but not both. Interstate pipelines are located through more than one state and fall under federal jurisdiction and classification. Intrastate pipelines are located in a single state, and different states may classify the pipeline differently. Procedures should be relevant to the states involved.

10.2.0 Abnormal Operating Conditions

Abnormal operating conditions (AOCs) are specified by federal law in *49 CFR Part 195.503 and CFR 192.803*. The Operator Qualification Rule requires that an individual recognize and respond to any AOC that could happen while performing a task. An AOC refers to any condition identified by the individual that may indicate failure of a component, or deviation from normal operations that may indicate a condition exceeding design limits or that may result in a hazard to persons, property, or the environment. The following list identifies the AOCs that can occur when operating a pipeline system:

- Abnormal facility condition
- Activation of a safety device
- Communications failure
- Control system failure
- Power interruption
- Fire, explosion, or natural disaster
- Pipeline system damage
- Unexpected hazardous liquid or carbon dioxide encountered
- Unexplained flow rate deviation
- Unexplained pressure deviations
- Unexplained status change
- Other company specific circumstances

10.3.0 Company Policies and Procedures

Company policy and procedure manuals typically include:

- Operating procedures manual
- Maintenance procedures manual
- Safety procedures manual
- Emergency procedures manual

Each procedure will typically include procedure details, purpose, responsibilities, and other supporting information. There are many different requirements depending on the company and the location of the pipeline.

10.4.0 Qualifications

With the introduction of the new Operator Qualification Rule in *49 CFR 195.503*, pipeline operators are now required to have a written qualification program. Required elements of the qualification program include:

- Identification of covered tasks
- Qualification in the performance of covered tasks through operator evaluations
- Evaluation measures for individuals following a DOT-reportable accident or for other reasonable causes
- Communication processes
- Identification of intervals at which reevaluations will occur

10.5.0 Documentation

Documentation is an important part of the pipeline operation from origin to end point. The DOT requires that many operations and maintenance procedures be documented to record their correct and timely performance. Problems must be documented to track down the cause and to prevent recurrence.

Some documents are forms that must be completed by operations personnel and technicians. Others are forms, reports, and summaries that are printed out periodically through the control system. Some of the documentation required for operations includes:

- Accident/incident reports
- Operations logs
- Work orders
- Daily events/event logs
- Personnel qualifications
- Inspection and repairs

10.5.1 Accident/Incident Reports

Accident/incident reports are required, as specified in the Office Pipeline System (OPS) regulations, whenever there is a reportable incident along the pipeline. An incident is defined as an unexpected situation that causes, or could be expected to cause, an operational problem, harm to personnel, or damage to the environment. Reportable situations are defined in the regulations.

10.5.2 Operations Logs

Operations logs record the daily events of operation. These are usually summarized on reports printed from the control system; they include all activities communicated to and from the control room as associated with the pipeline.

10.5.3 Work Orders

Work orders and/or work permits are issued when specific maintenance or other tasks need to be performed. Each work order or permit has its own number and is recorded on a work order or permit log along with a description of the job and other details of the assignment.

These forms are required to ensure systematic maintenance of the pipeline, to assist with scheduling, and to record work done. Work orders and schedules for tasks done at specific time intervals can be triggered by the computer system. Not all pipeline companies use a work order system.

10.5.4 Daily Events and Event Logs

Daily events are recorded in the event log as specific tasks or activities that are performed during the day's operations. These might include repairs, inspections, AOCs, investigations, maintenance, or many other events.

10.5.5 Personnel Qualifications

The OPS regulation provides for formal documentation of the qualifications of personnel responsible for pipeline operations and maintenance in each company. The regulation was designed to reduce the risk of accidents on pipeline facilities caused by human error. Each company must have a written program specific to its own system for the qualification of individuals, including employees, contractors, and subcontractors who work on that system. These qualifications often cover about fifty different safety-related tasks, usually by position, such as pipeline controllers, and corrosion, maintenance, and operations personnel. The rule requires that a number of specific elements be present in all operator plans, including:

- Personnel must demonstrate the ability to recognize and respond to abnormal operating conditions.
- Companies must keep records that demonstrate compliance to the OPS.

10.5.6 Inspection and Repairs

When repairs are made, the pipe specifications, defect description, stress levels, and action taken must be documented and maintained in case of an audit. Other information, in accordance with the requirements of regulatory authorities, must also be prepared. Some documents are forms that must be filled in by operations and maintenance personnel. Others are forms, reports, and summaries that are printed out periodically through the control system.

Federal guidelines require that all inspections, treatments, and repairs be documented. Some of the documentation required for corrosion control includes:

- Cathodic protection system inspections
- In-line inspection results
- Pipeline coating inspections
- Incident reports
- Maintenance reports

Record-keeping is a huge and important task. Tremendous amounts of inspection and monitoring data are generated over the life of a pipeline, including:

- Development of written pipeline integrity procedures and coordination with operation, maintenance, contingency, and emergency response procedures
- Identification and implementation of repairs and modifications
- Program evaluation

Many software programs are now available to make record-keeping easier and to help companies stay in compliance with regulations. Failure to maintain good documentation can result in significant penalties for your company.

Summary

The term *pipeline* has many meanings. In the simplest terms, it applies to the pipe and fittings used to construct facilities for transporting commodities. In a broader sense, it applies to all of the facilities in addition to the pipe and fittings that are needed to transport the commodities from the point of origin to the point of delivery. Pipeline facilities include the means of measuring, controlling, and accounting for quantities and qualities of product delivered from the custody of the original transporter to the shipper.

In order to efficiently operate and maintain the pipeline facilities, personnel must understand the basic principles of pipeline hydraulics when working with liquid pipelines, and pipeline pneumatics when working with gas pipelines. Personnel must also familiarize themselves with basic pipeline equipment and its proper use.

Electrical power systems supply power to the pipeline equipment and facilities. Electrical and instrumentation technicians maintain and repair this equipment to ensure that the pipeline remains in service. Another major task of pipeline personnel is to monitor corrosion and practice measures to prevent or minimize the presence of corrosion on the pipeline and equipment.

Each pipeline company has published operations and maintenance and emergency procedures. These procedures are based on federal, state, and local regulations and on industry standards. All pipeline personnel must familiarize themselves with these standards in order to operate the pipeline in a manner that will ensure the safety of the employees and the general public and protect the environment.

Review Questions

1. Oil and gas are formed from _____.
   a. decaying organic matter
   b. buried rocks
   c. salt domes
   d. groundwater

2. Different types of pumps or valves are used to cap the well, depending on the _____ of the fluid being withdrawn.
   a. temperature
   b. destination
   c. properties
   d. amount

3. Lines from the oil fields are called _____ lines.
   a. field
   b. main
   c. distribution
   d. gathering

4. The pipe must be _____ tested to verify the pipeline integrity.
   a. strength-
   b. X-ray-
   c. hydrostatically
   d. sonar-

5. Different petroleum products can be shipped consecutively in the same pipeline. This is called _____.
   a. bunching
   b. mixing
   c. batching
   d. blending

6. Large volumes of natural gas are typically stored in _____.
   a. underground storage fields
   b. surge tanks
   c. relief tanks
   d. open roof tanks

7. Oil containing a large amount of sulfur compounds is called a _____.
   a. sweet crude
   b. sour crude
   c. paraffin
   d. light crude

8. There is a set of _____ for each hazardous substance, listing its properties and hazards.
   a. chemical descriptions
   b. OSHA labels
   c. product papers
   d. MSDSs

9. The marine environment is particularly sensitive to _____.
   a. high pressure
   b. overproduction
   c. oil spills
   d. vapor accumulation

10. A(n) _____ depicts the functional layout of a piping system and its components as clearly and accurately as possible.
    a. P&ID
    b. alignment sheet
    c. strip map
    d. blueprint

11. _____ is the software used to monitor and control the flow of millions of barrels of oil and products through the pipeline.
    a. RTUS
    b. SCADA
    c. SEACA
    d. PLCS

12. The science of the behavior of fluids at rest or in motion is referred to as _____.
    a. pneumatics
    b. hydraulics
    c. compressibility
    d. viscosity

13. Boyle's law states that when temperature is kept constant and volume is decreased, pressure will _____.
    a. decrease
    b. increase
    c. stay the same
    d. not be affected

14. In a typical positive-displacement pump, a _____ slides back and forth in a cylinder.
    a. rotor
    b. vane
    c. piston
    d. blade

15. _____ are used to control the rate of flow in a line, to open or shut off a line, or to serve as automatic safety devices.

a. Motors
b. Valves
c. Pumps
d. Manifolds

16. Tanks must be calibrated for _____.

a. strength
b. sealing ability
c. capacity
d. wall thickness

17. _____ are used to join three pipe ends.

a. Flanges
b. Unions
c. Elbows
d. Tees

18. The pipeline uses _____ to control the flow path of oil and petroleum products.

a. manifolds
b. pumps
c. tanks
d. vents

19. Various _____ are used along the pipeline to monitor the volume of fluid.

a. gauges
b. tests
c. meters
d. provers

20. Communications failure is one example of an _____.

a. emergency
b. AOC
c. accident
d. interruption

GLOSSARY

Trade Terms Introduced in This Module

Anomaly: A defect in the pipeline or pipeline coating caused by corrosion, wear, or other external means.

Barrel: The common measure of petroleum products: one barrel equals 42 U.S. gallons or 0.15899 cubic meters or 9,702 cubic inches.

Cathodic protection: A means of preventing the destructive electrochemical process of corrosion of metal by using the metal as the cathode of a cell with a sacrificial anode.

Corrosion: The gradual destruction, dissolving, or wearing away of a substance, usually a metal or an alloy, caused by a chemical or electrochemical reaction with its environment. Corrosion can be a problem in different locations along the pipeline, both externally and internally.

Crude oil: Unrefined liquid petroleum. Crude oil ranges in color from yellow to black, and may have a paraffin, asphalt, or mixed base. Crude oils may be referred to as heavy or light according to API gravity, with the lighter oils having the higher gravities.

Cubicle: An enclosure for high-voltage equipment.

Department of Transportation (DOT): The federal agency that sets standards for, and regulates, transportation.

Hydrostatic testing: The most common final quality check on a complete pipeline. Water is pumped into the line and pressured up to a specified level. The pipe is checked for leaks or ruptures.

Main line: The primary pipe between distant points; it is also called the trunk line.

Manifold: An accessory system of piping to a main piping system that serves to divide the flow into parts, to combine several flows into one, or to reroute a flow to several possible locations.

Material safety data sheet (MSDS): A form containing information on a potentially hazardous material, including names, contents, test results, characteristics, potential hazards, and handling procedures.

Natural gas: A highly compressible, highly expandable mixture of hydrocarbons with a low specific gravity. It occurs naturally in a gaseous form. Besides hydrocarbon gases, it may contain nitrogen, helium, carbon dioxide, hydrogen sulfide, and water vapor. The gases are variable in form and may be found as gases or liquids under various conditions of pressure and temperature. Natural gas liquids include butane, propane, and natural gasoline.

Natural gas liquids (NGLs): The portions of reservoir gas that are liquefied at the surface in lease separators, field facilities, or gas plants.

Petroleum: Unrefined liquid hydrocarbons, including *natural gas liquids (NGLs)*.

Pig: A scraping tool with blades or brushes that is forced through a pipeline to clean out accumulations of wax, scale, and debris from the pipe walls; it is also called a line scraper or go-devil.

Product: Any fluid, refined or not, that is transported through a pipeline.

Refined products: Manufactured petroleum products generated from crude oil.

Refinery: System of process units used to convert crude petroleum into fuels, lubricants, and other petroleum-derived products.

Seismic activity: The phenomena of earth movements, earthquakes, and earth vibrations, including those that are artificially induced.

Tank: A metal, plastic, or wooden container used to store a liquid. Three types include mud tanks for drilling, production tanks, and storage tanks.

Throughput: The volume of product that goes through a section of equipment during a specified amount of time.

The NCCER makes every effort to keep these textbooks up-to-date and free of technical errors. We appreciate your help in this process. If you have an idea for improving this textbook, or if you find an error, a typographical mistake, or an inaccuracy in NCCER's Contren™ textbooks, please write us, using this form or a photocopy. Be sure to include the exact module number, page number, a detailed description, and the correction, if applicable. Your input will be brought to the attention of the Technical Review Committee. Thank you for your assistance.

Instructors – If you found that additional materials were necessary in order to teach this module effectively, please let us know so that we may include them in the Equipment/Materials list in the Instructor's Guide.

Write: Curriculum Revision and Development Department
National Center for Construction Education and Research
P.O. Box 141104, Gainesville, FL 32614-1104

Fax: 352-334-0932

E-mail: curriculum@nccer.org

Craft ____________________ Module Name ____________________

Copyright Date ____________ Module Number ____________ Page Number(s) ____________

Description

____

____

____

____

(Optional) Correction

____

____

____

(Optional) Your Name and Address

____

____

____

Module 67102-02

Basic Pipeline Pneumatics and Equipment

Introduction to the Pipeline Operator Qualification Rule, API, and NCCER

It could be said that petroleum products keep America running. Petroleum products like oil and natural gas are the principal sources of energy in the United States and are used in many manufacturing and chemical processes. Petroleum products are primarily transported by pipeline. There are more than a million miles of liquid and natural gas pipeline in this country, ranging in diameter from two inches to four feet. These pipelines bring crude oil and natural gas from production fields to processing plants and refineries—and ultimately to the consumer.

Pipeline Operator Qualification

Petroleum products are considered hazardous, so pipeline personnel must take extra precautions and follow specific procedures when working with them. To help ensure the safe handling and transportation of petroleum products, the United States Department of Transportation (DOT) mandates safe practices as federal regulations.

One such regulation, the Operator Qualification (OQ) rule, requires pipeline operators to develop a written qualification plan to evaluate their personnel and their contract personnel (see *49 CFR Part 192 Subpart N—Gas*, and *49 CFR Part 195 Subpart G—Liquid*). The evaluation hinges on the operator's ability to perform certain covered tasks and to be able to recognize and respond to abnormal operating conditions that may be encountered while performing these covered tasks.

Covered Tasks

A pipeline performance task is deemed *covered* if the activity passes the following four-part test:

1. Is it performed on a pipeline facility?
2. Is it an operations or maintenance task?
3. Is it performed as a requirement of *49 CFR Part 195 (Liquid)* or *49 CFR Part 192 (Gas)*?
4. Does it affect the operation or integrity of the pipeline facility?

If the answer to all four questions is *yes*, then the activity is considered *covered* and must figure into the evaluation process of pipeline personnel.

American Petroleum Institute

The OQ rule and its related covered tasks provided the catalyst to develop a unified pipeline training and assessment program—so that pipeline operators and owners have ready access to the assessment and training tools they need. This effort was the direct result of the vision and leadership provided by The American Petroleum Institute (API) and API members.

Efforts to unify the industry are nothing new to API. Indeed, the impetus for forming the Institute in 1919 was the need to standardize engineering specifications for drilling and production equipment. Today, the most pressing issues revolve around public perceptions and government policies toward the industry. Speaking with one voice on these issues has become as essential as having interchangeable parts in the field. API's members determine the petroleum industry's positions on public policy as well as the standards that govern its day-to-day operations worldwide.

Consortium on Operator Qualification

API assembled a focus group whose primary purpose was to identify covered tasks and provide compliance recommendations to pipeline operators.

The Consortium on Operator Qualification (COOQ) became the guiding force for the identification and development of covered tasks for the liquid pipeline industry. Their work product, API publication 1161, *Guidance Document for the Qualification of Liquid Pipeline Personnel*, formed the basis for the resulting Pipeliner Training and Assessment Program.

The covered task list created in API publication 1161 has now evolved to include gas covered tasks and is maintained by NCCER. The most current covered task list can be downloaded at www.nccer.org.

National Center for Construction Education and Research

With the guidance and support of API, its members, and the Consortium on Operator Qualification, the National Center for Construction Education and Research (NCCER) was able to design and implement the Pipeliner Training and Assessment Program.

NCCER is a not-for-profit 501(c)(3) education foundation established in 1995 by the world's largest and most progressive construction and maintenance contractors and national contractor associations. NCCER was founded to address the

severe workforce shortage facing the construction and maintenance industries and to develop a standardized training process and curricula. Today, NCCER is supported by hundreds of leading construction and maintenance companies, manufacturers, and national associations and developed more than 35 construction and maintenance titles.

NCCER also maintains a National Registry, which contains the records of trainees and instructors who have received NCCER training. All completion information for every trainee is sent to the NCCER and filed within the Registry. The Registry can then confirm training and skills for workers as they move from company to company, state to state, or even within their own company.

Standardized Curriculum (NCCER's Contren™ Learning Series)

NCCER curricula are developed using the expertise of industry subject matter experts (SMEs). The use of SMEs as an integral part of every new development or revision ensures that the curriculum is current, industry-driven, and indicative of industry practices. The development and/or revision process is facilitated from start to finish by NCCER project managers, who work closely with the SMEs and a select team of technical writer(s) on the development and presentation of the curriculum.

Pipeline: Five Training Paths

The pipeline curriculum focuses on essential skills and competencies within five skill areas:

- Corrosion Control
- Electrical and Instrumentation
- Maintenance
- Mechanical
- Operations (Field and Control Center)

SMEs from the pipeline industry have determined the skills and competencies needed based on API's covered tasks and the consensus of the group/industry. Specific company information is generally not incorporated into the curriculum unless the SMEs determine that the information is the industry standard.

Training Modules and Covered Tasks

The smallest unit of instruction in the curriculum is a *module*. Each curriculum is composed of modules (the equivalent of stand-alone chapters) that are logically sequenced into complete *levels* of instruction. Every module is task-oriented and requires the successful completion of a written (knowledge-based) examination and a performance (hands-on) test. Modules may be individually selected and/or combined for efficient task training. Modules in the pipeline curricula, to the extent possible, will align with API-developed covered tasks. While not all modules will relate to covered tasks, all covered tasks will relate to at least one module.

Benefits of Training to the Pipeline Marketplace

Applying NCCER's standardized training process allows pipeline owners and contractors in the United States to establish the benchmark for acceptable skill and competency levels within the pipeline industry. This process will assist training organizations in developing and/or improving standardized pipeline training programs that effectively match their resources and operating conditions while responding to regulatory demands and to the needs of the industry. The process will also recognize training organizations that meet the minimum standards of competence, warranting regulatory, public, and professional confidence. This in turn will foster national unity within the pipeline industry for the mutual benefit of individual technicians, owners, contractors, and ultimately pipeline-users and society.

Additional Resources

To review the Pipeliner Training and Assessment Program Implementation and Resource Guide and the latest version of pipeline covered tasks, visit NCCER's Web site at www.nccer.org, or contact:

National Center for Construction Education and Research
PO Box 141104
Gainesville, FL 32614-1104
(888) 622-3720
Web site: www.nccer.org

For more information on API, please contact:

American Petroleum Institute
1220 L Street, Northwest
Washington, D.C. 20005
(202) 682-8000
Web site: www.api.org

To order *API Pub 1161*, contact Global Engineering Documents by calling 1-800-854-7149 or visiting their Web site at www.global.his.com.

Additional sources of information on the pipeline operator qualification regulation include:

DOT – Office of Pipeline Safety
www.ops.dot.gov

Transportation Safety Institute
www.tsi.dot.gov

COURSE MAP

This course map shows all of the modules in the Gas Pipeline Operations curriculum. The suggested training order begins at the bottom and proceeds up. Skill levels increase as you advance on the course map. The local Training Program Sponsor may adjust the training order.

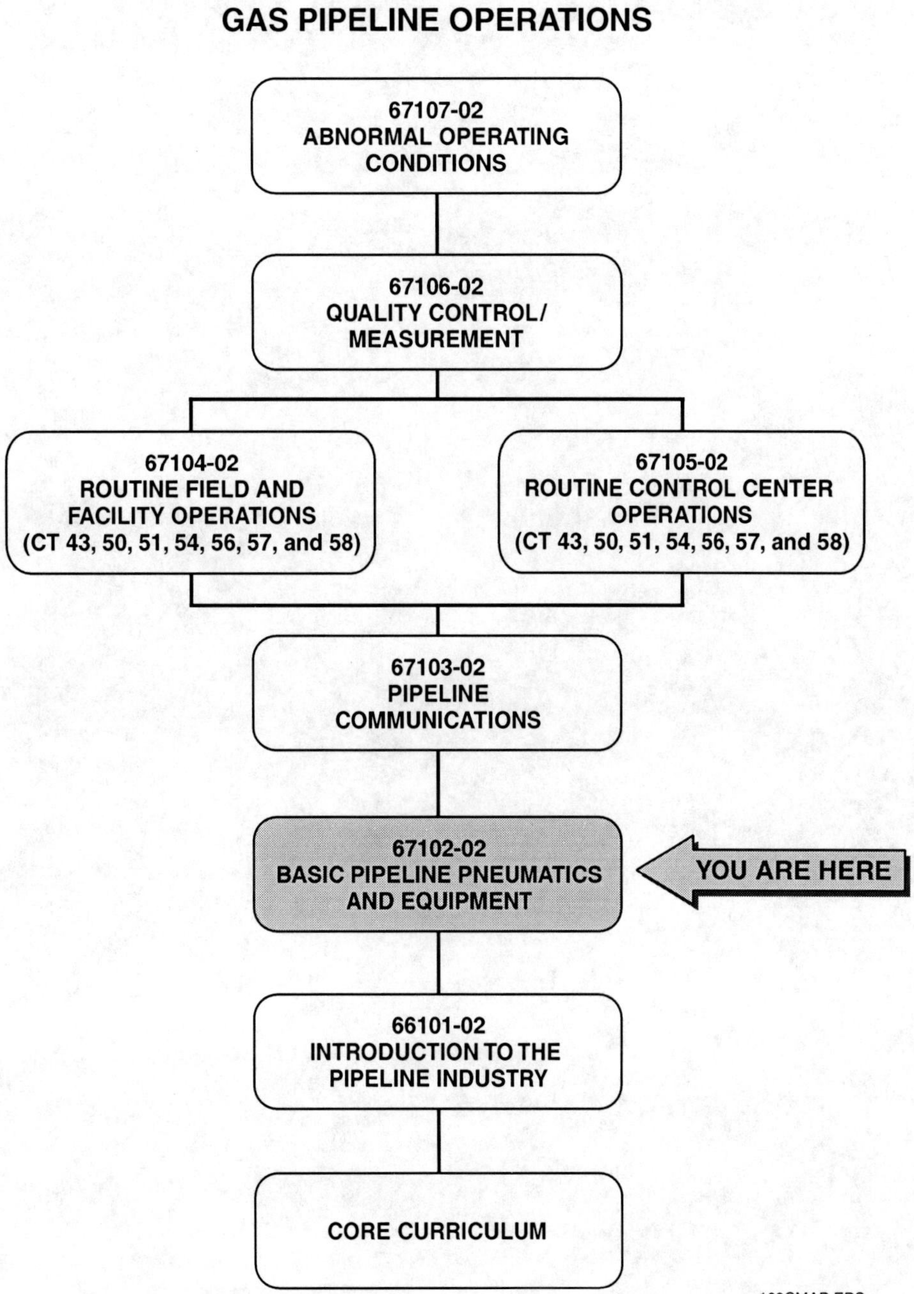

102CMAP.EPS

MODULE 67102-02 CONTENTS

Figures

Tables

MODULE 67102-02

Basic Pipeline Pneumatics and Equipment

Objectives

When you have completed this module, you will be able to do the following:

1. Explain pneumatic safety.
2. Explain the physical characteristics of gas.
3. Explain compressing gases.
4. Explain the pneumatic transmission of energy.
5. Explain the principles of compressor operation.
6. Identify and explain types of gas pipeline equipment.
7. Identify and explain pneumatic system components and symbols.
8. Demonstrate an understanding of the design limits of pipelines.

Prerequisites

Before you begin this module, it is recommended that you successfully complete the following: Core Curriculum; Pipeline Core, Module 66101-02.

1.0.0 ◆ INTRODUCTION

Pneumatic systems use a gas, usually air, under pressure to create energy and motion. The earliest and simplest type of pneumatic tool is the bellows. Air is compressed inside the accordion-shaped bellows and forced out at greater speed. A **compressor,** which is also a pneumatic device, is a more complex version of a bellows. Compressors are used on the gas pipeline to boost the speed of flowing gas.

When natural gas is released from an underground well, it is under a lot of pressure, which causes the gas to flow rapidly into a pipeline; however, as the gas flows through the pipeline, friction slows its movement, and the pressure drops. Compressors are used to bring the gas pressure back up and keep the gas moving. Because gas pipelines cover thousands of miles, many compressor stations are located along the pipeline. The gas gets a pressure boost as it passes through these stations. Pipeline personnel must carefully calculate compression requirements to properly size the compressors and determine the number and spacing of compressors along the line.

This module introduces you to the basics of working with pneumatic systems and equipment. You will learn about some of the science that governs how gases behave under different conditions. You will also be introduced to some of the equipment used on the gas pipeline and acquire a basic understanding of pipeline design.

2.0.0 ◆ PNEUMATIC SAFETY

Any element placed under pressure develops a certain amount of force. This force is what moves gas through a pipeline. However, if not properly handled, this force can be dangerous. Even low-pressure systems can cause injuries or damage equipment. In addition, working around pneumatic equipment may expose you to toxic or flammable gases. Your company has specific safety procedures for working around pneumatic equipment. You must study these procedures and follow them to ensure your safety and the safety of your co-workers. You must understand and follow all safety procedures before attempting to troubleshoot or work on pneumatic equipment or systems. A list of general safety guidelines follows:

- Always wear the appropriate personal protective equipment.
- Learn and follow company safety procedures.

- Study the **material safety data sheet (MSDS)** if appropriate.
- Study the appropriate technical manuals for proper procedures and specific safety guidelines.
- Ensure that the equipment or system is **locked out** and **tagged out.**
- Ensure that the equipment or system is de-energized both electrically and mechanically.
- Check pressure gauges to ensure that the pressure reading is at zero. Be aware that pressure may be trapped in the equipment or the system.
- Study the possible hazards before opening or closing a **valve** mechanically or electrically.
- Ensure that machines are in a down or safe position.
- Run breathing air through a dehydrator and a carbon dioxide monitor.
- Check the area with a gas detection instrument to ensure that oxygen has not been displaced by another gas.

> **NOTE**
> Gas detection instruments are available in a wide variety of styles and applications. Some are programmed to test for a single gas. Others can continuously monitor for several gases at one time. There are both portable units for fieldwork and fixed units for indoor installations. Many include audible and visual alarms.

- Replace defective parts with identical parts. Pay special attention to sizes and pressure ratings.
- Check your work to ensure that all repairs have been completed properly.

3.0.0 ◆ CHARACTERISTICS OF NATURAL GAS

Natural gas is a hydrocarbon mixture. The main gases found in this mixture are methane and ethane. Other hydrocarbon gases in the mixture include propane, butane, pentane, and hexane. Natural gas may also contain some nonhydrocarbon gases, such as nitrogen, carbon dioxide, hydrogen, and hydrogen sulfide. Some water is also present in natural gas.

The main ingredient of natural gas—methane—is colorless, odorless, and flammable. Some natural gas can be used without any processing or refinement. However, most natural gas is processed to remove unwanted elements, such as water vapor, carbon dioxide, and hydrogen sulfide. The resulting refined gas is almost totally methane. Natural gas may be transported through **gathering lines, transmission lines,** or **distribution lines.**

Like any other substance, gas is made up of molecules. Gas molecules do not stick together like molecules in some other substances. For example, if you pour water from a container, the water molecules stick together to form puddles. Gas molecules, however, scatter or disperse into the atmosphere. When gas is contained in a pipeline, the molecules race along, constantly colliding with each other and with the walls of the pipeline as they try to disperse. A pressure gauge interprets these billions of collisions as a single pressure reading. All that activity produces heat, and the greater the compression, the greater the heat. The hotter the gas is, the faster the molecules move and the more heat energy is generated. However, the heat cannot be allowed to build indefinitely. Gas flowing through the pipeline must be diverted to cooling towers where its temperature is brought down to acceptable levels before it is further transported. This is done to protect the pipeline from damage.

3.1.0 Compression

When a gas fills an open container, the pressure and temperature inside the container is the same as the pressure and temperature outside of it. If the container has a close-fitting piston, gas volume is reduced, or compressed, when the piston is pushed in. As the gas is compressed, the speed and temperature of the gas molecules increase. Pressure also increases because the same number of molecules occupies a smaller space, and the molecular collisions are more frequent.

Pipeline personnel must control three variables—pressure, volume, and temperature—to compress and move gas through a pipeline. These variables have a close relationship to one another. If one variable remains constant, the other two change either directly or inversely (opposite to each other). The relationships between these variables are explained with three gas laws: Boyle's law, Amontons' (Gay-Lussac's) law, and Charles' law. These gas laws apply to any gaseous element.

3.2.0 Boyle's Law

Boyle's law was developed by Irish chemist Robert Boyle in 1662. This law explains the relationship between pressure and volume if temperature remains constant. If volume decreases, pressure increases. If volume increases, pressure decreases. This is an example of an inverse relationship; when one value goes up, the other value always goes down.

Following is an example of how Boyle's law works in a pipeline when temperature is constant. When volume is greater than pressure, gas molecules have farther to travel, so they hit the pipeline walls less often. Gas pressure then decreases because there are fewer molecule hits over time. When volume is less than pressure, gas molecules have a shorter distance to travel. Therefore, they strike the pipeline walls more often. Pressure therefore increases because there are more molecule hits over time.

3.3.0 Amontons' (Gay-Lussac's) Law

Amontons' law was developed by French physicist Guillaume Amontons in 1699. Because Joseph Gay-Lussac built the apparatus to prove the theory, this law is sometimes given his name. Amontons' law is an example of a direct relationship. When there are two connected values, whatever happens to one value also happens to the other. This law explains the relationship between pressure and temperature when volume remains constant. When temperature increases, pressure also increases. If temperature decreases, pressure also decreases.

Following is an example of how Amontons' law works in a pipeline when volume remains constant. When the temperature increases, gas molecules move faster. Faster-moving molecules hit the pipeline walls more often, causing the gas pressure inside the pipeline to increase. When the temperature decreases, gas molecules move more slowly. Slower-moving molecules hit the pipeline walls less often, causing the pressure inside the pipeline to decrease.

3.4.0 Charles' Law

Charles' law was developed by French scientist Jacques Charles in 1787. This law also defines a direct relationship. It gives the relationship between volume and temperature when pressure remains constant. When volume increases, temperature increases. If volume decreases, temperature decreases.

Following is an example of how Charles' law works in a pipeline when pressure remains constant. When temperature increases, gas molecules move faster, hitting the pipeline walls more often. The gas pressure inside the pipeline increases as well, but only for an instant. The greater pressure on the inside of the pipeline walls pushes the walls outward, increasing the volume. When this happens, the gas molecules have to travel farther to hit the walls. The lower number of hits drops the pressure back to its constant value.

4.0.0 ◆ PNEUMATIC TRANSMISSION OF ENERGY

Pneumatic systems use compressed air or other gases to produce energy. Compressed air is commonly used to power such tools as impact wrenches and sandblasting equipment. Flexible hoses or tubes carry the compressed air from the compressor to the tool.

A compressed air transmission system carries air through pipes from the compressor to an air-powered device (see *Figure 1*). Most air transmission systems consist of a main line, branch lines, and feeder lines. The feeder lines connect to individual pieces of air-powered equipment. Sometimes air distribution is more efficient if the ends of the lines are tied together to form a loop system. This loop allows two-way airflow to a point of heavy demand.

In the pipeline industry, field personnel may use bottled high-pressure gas to charge relief valves and flow-control devices. Bottled gas may also be used to purge small sections of piping or small tanks. The gases used for these smaller jobs include nitrogen, argon, and helium.

The pressure of the gas transported in the pipeline can be reduced to open or close valves. The main-line compressors for gas pipeline compressor stations usually are started with compressed air.

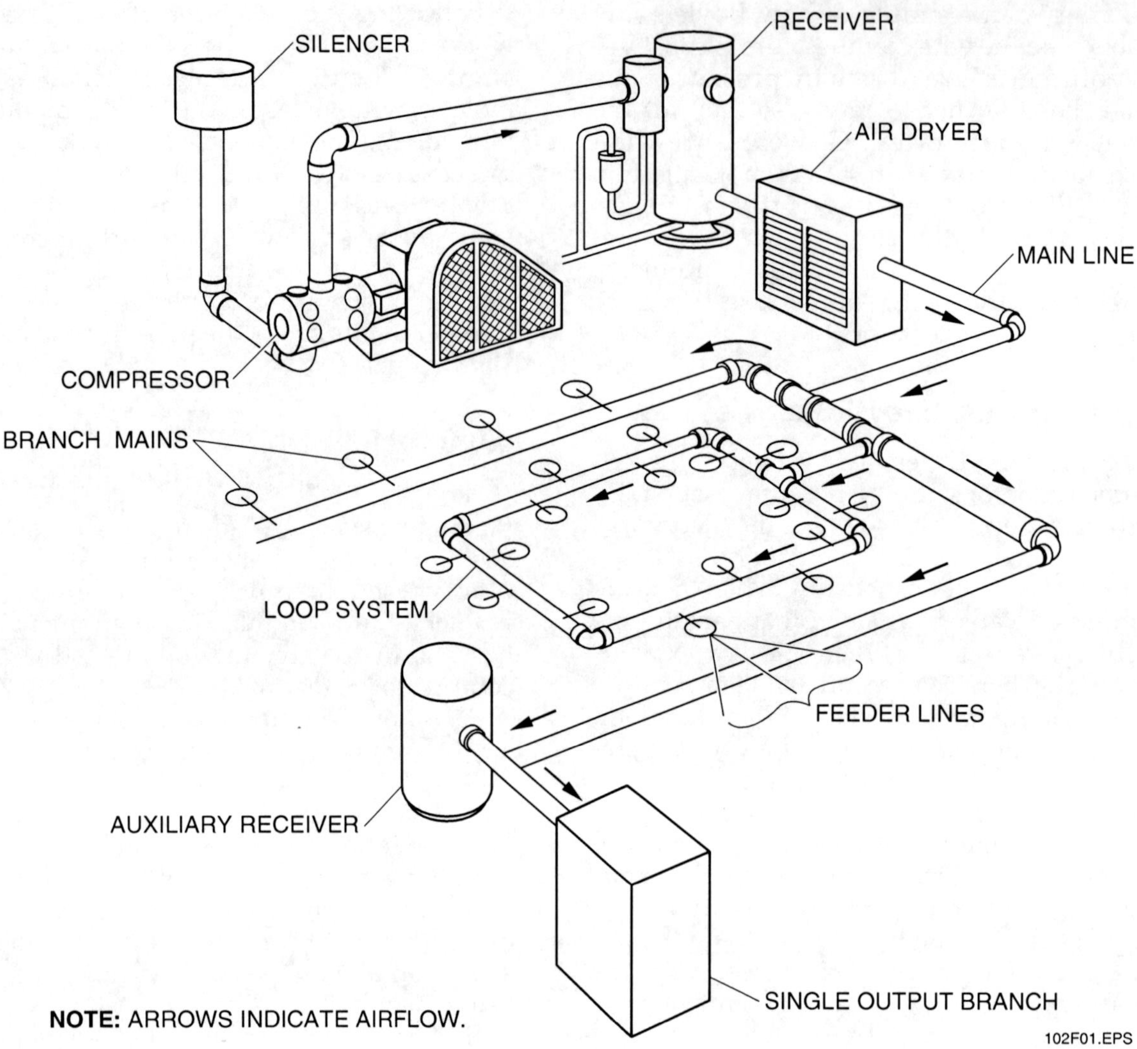

Figure 1 ◆ Compressed air transmission system.

Review Questions

Sections 2.0.0–4.0.0

1. Before attempting to troubleshoot or work on pneumatic equipment or systems, you must ensure that the equipment or system is _____.

a. locked out and tagged out
b. marked with a color-coded sign
c. marked with an authorized work order
d. redlined as a service item

2. The main gases found in natural gas are _____.

a. butane and propane
b. butane and ethane
c. methane and ethane
d. methane and pentane

3. As gas is compressed in a pipeline, it _____.

a. liquifies
b. separates into butane and ethane
c. cools off
d. heats up

4. _____ explains the relationship between pressure and volume if temperature remains constant.

a. Boyle's law
b. Amontons' law
c. Guy-Lussac's law
d. Charles' law

5. In the pipeline industry, field personnel may use _____ to charge relief valves and flow-control devices.

a. pressure-bleeding units
b. bottled low-pressure gas
c. bottled high-pressure gas
d. a pneumatic unit

5.0.0 ◆ COMPRESSORS

A compressor increases the pressure of gas by reducing its volume. Compressor stations are generally placed at approximately 100-mile intervals along transmission lines. Compressors may be located at more frequent intervals along distribution lines. Some compressors are powered by natural gas from the pipeline, and some are powered by electrically operated motors.

Every compressor is made up of one or more basic components. A single-stage compressor may be made up of a single component or a group of components. Single-stage compressors may not be able to handle all operating conditions. For example, too great a **compression ratio** may raise the **discharge temperature** too much. To solve this problem, components are combined to form a multistage unit with two or more stages of compression. The gas is cooled between the stages to reduce its temperature and volume before it enters the next stage. Each stage of a multistage unit contains an individual compressor. Each compressor is sized to operate with the other compressors in the series, and all of the compressors in the series operate from one power source.

The two main types of compressors are positive-displacement and continuous-flow. There are several different versions of compressors within these two main types (see *Figure 2*).

5.1.0 Positive-Displacement Compressors

In positive-displacement compressors, air is forced into an enclosed space. Pistons, gears, or pumps then increase the pressure within that space. Positive-displacement compressors are classified as either reciprocating or rotary.

In the reciprocating compressor, a piston compresses the air (see *Figure 3*). The piston rests in a cylinder that has both **suction** and **discharge** valves. The term *suction* (sometimes called the inlet or intake) refers to the gas flow coming into the compressor. The term *discharge* (sometimes called the outlet) refers to the gas flow leaving the compressor. The suction valve opens when the pressure in the cylinder is slightly below the **suction pressure.** The discharge valve opens when the pressure in the cylinder is slightly above the **discharge pressure.** This type of compressor may be single acting or double acting. In the single-acting unit, gas is compressed on only one side of the piston. In the double-acting unit, gas is compressed first on one side of the piston and then the other.

In the rotary compressor, gears or an electric pump move confined air to smaller and smaller spaces. As the volume of air decreases, the pressure increases. It is a simpler unit than the reciprocating compressor. There are no valves or pistons. Instead, rotors that fit together like gears compress air flowing through the unit. Following are common types of rotary compressors:

- Sliding-vane compressors
- Liquid-piston compressors
- Twin-impeller, straight-lobe compressors
- Helical, or spiral-lobe, compressors

5.1.1 Sliding-Vane Compressors

A sliding-vane compressor consists of a case that holds a rotor assembly and vanes. The compression cycle of a sliding-vane compressor is shown in *Figure 4*. Gas flows in through the suction port. The rotor turns and traps the gas in pockets formed by the vanes. Once the pockets between the vanes are filled, the suction port closes. The pocket volume decreases as the rotor turns, and the gas is compressed. The compressed gas is then pushed out of the discharge port.

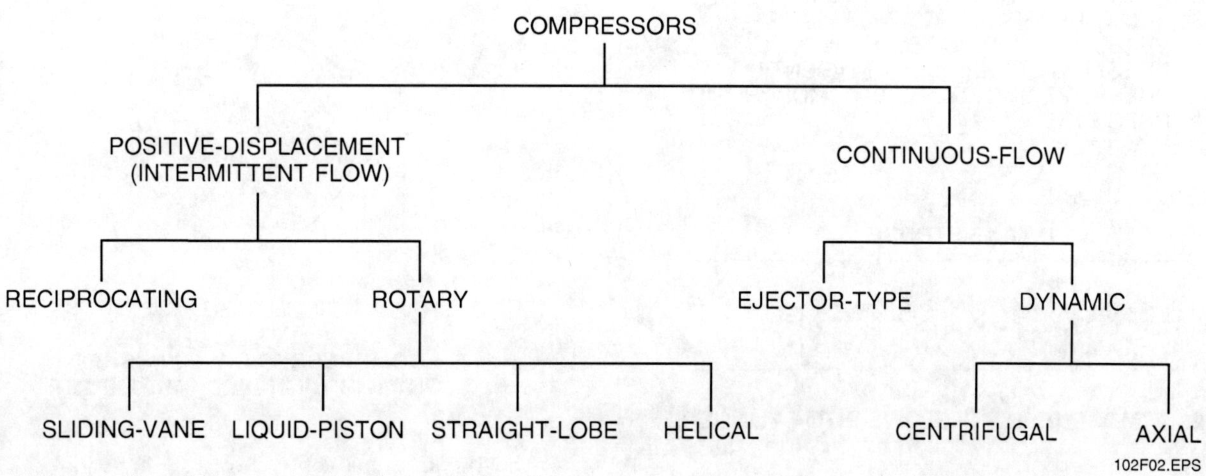

Figure 2 ◆ Various types of compressors.

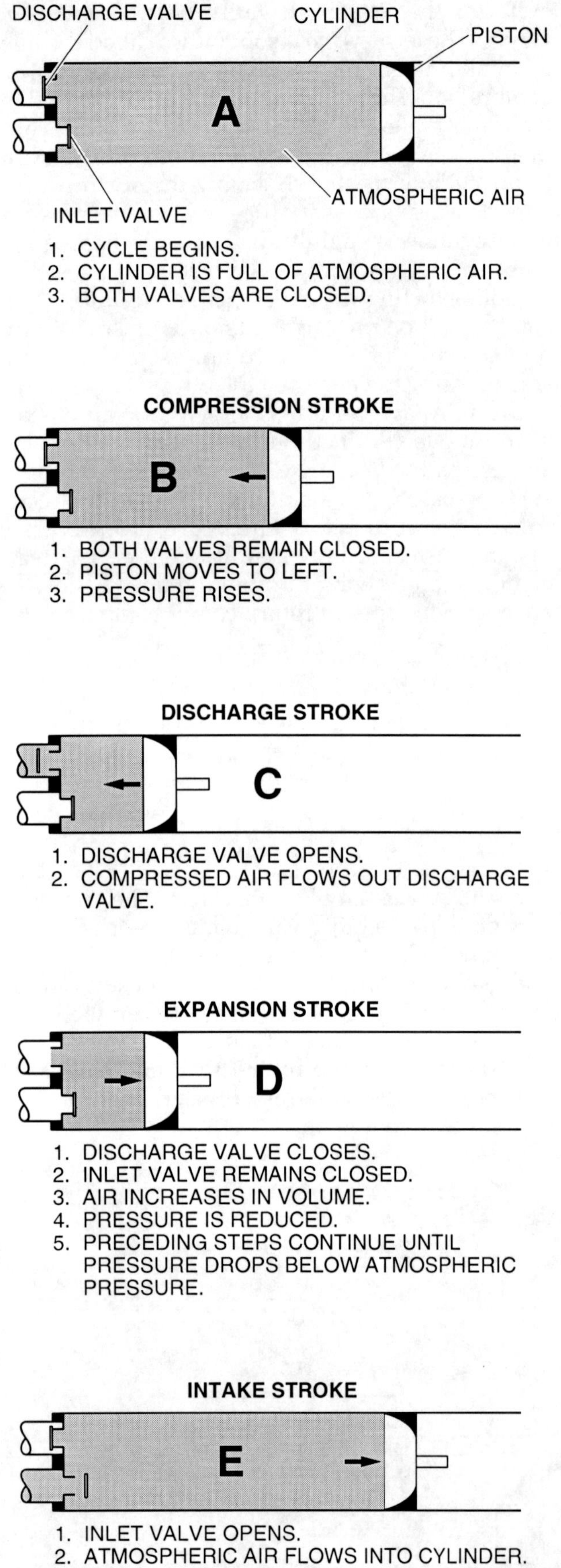

Figure 3 ◆ Steps in a reciprocating compressor cycle.

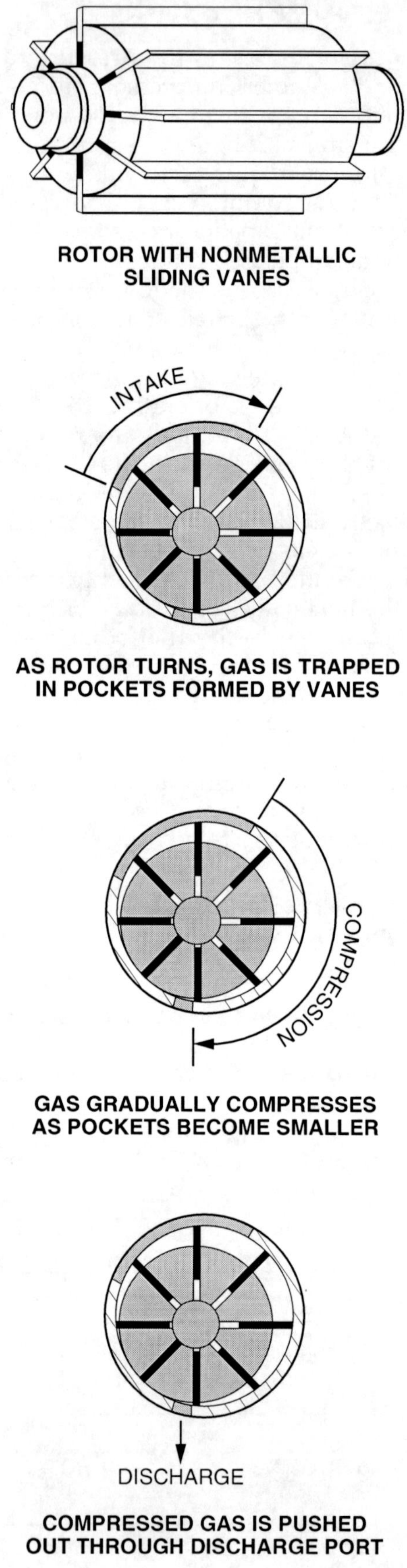

Figure 4 ◆ Compression cycle of a sliding-vane compressor.

5.1.2 Liquid-Piston Compressors

A liquid-piston, or liquid-ring, compressor uses a rotor with forward-turned blades. These blades turn around a central cone that holds suction and discharge ports (see *Figure 5*). The blades drive a ring of liquid compressant around the inside of the casing. The liquid acts as a lubricant, sealer, and coolant. Some liquid is trapped in the space between the blades, and as the rotor turns, it moves in and out of the space, creating a liquid piston. Lubrication is required only in the bearings located outside the casing for this type of compressor; therefore, the compressed gas is oil-free.

5.1.3 Twin-Impeller, Straight-Lobe Compressors

The casing of the twin-impeller, straight-lobe compressor holds two identical impellers, or rotors. When viewed in cross section, each impeller looks like a figure eight (see *Figure 6*). The impellers rotate in opposite directions and mesh somewhat like gears. As they rotate, the impellers compress the gas from the suction port against the casing. External timing gears keep the impellers moving smoothly against one another.

5.1.4 Helical (Spiral-Lobe) Compressors

The casing of the helical, or spiral-lobe, compressor holds two rotors: a driven rotor and a gate rotor (see *Figure 7*). This unit compresses gas between the rotor lobes and the housing chamber. Unlike the lobes on the twin-impeller compressor, the lobes of the helical compressor are not identical. The male, or driven, rotor fits into a pocket formed by the female, or gate, rotor. In addition, the driven rotor uses about 90 percent of the power in this unit.

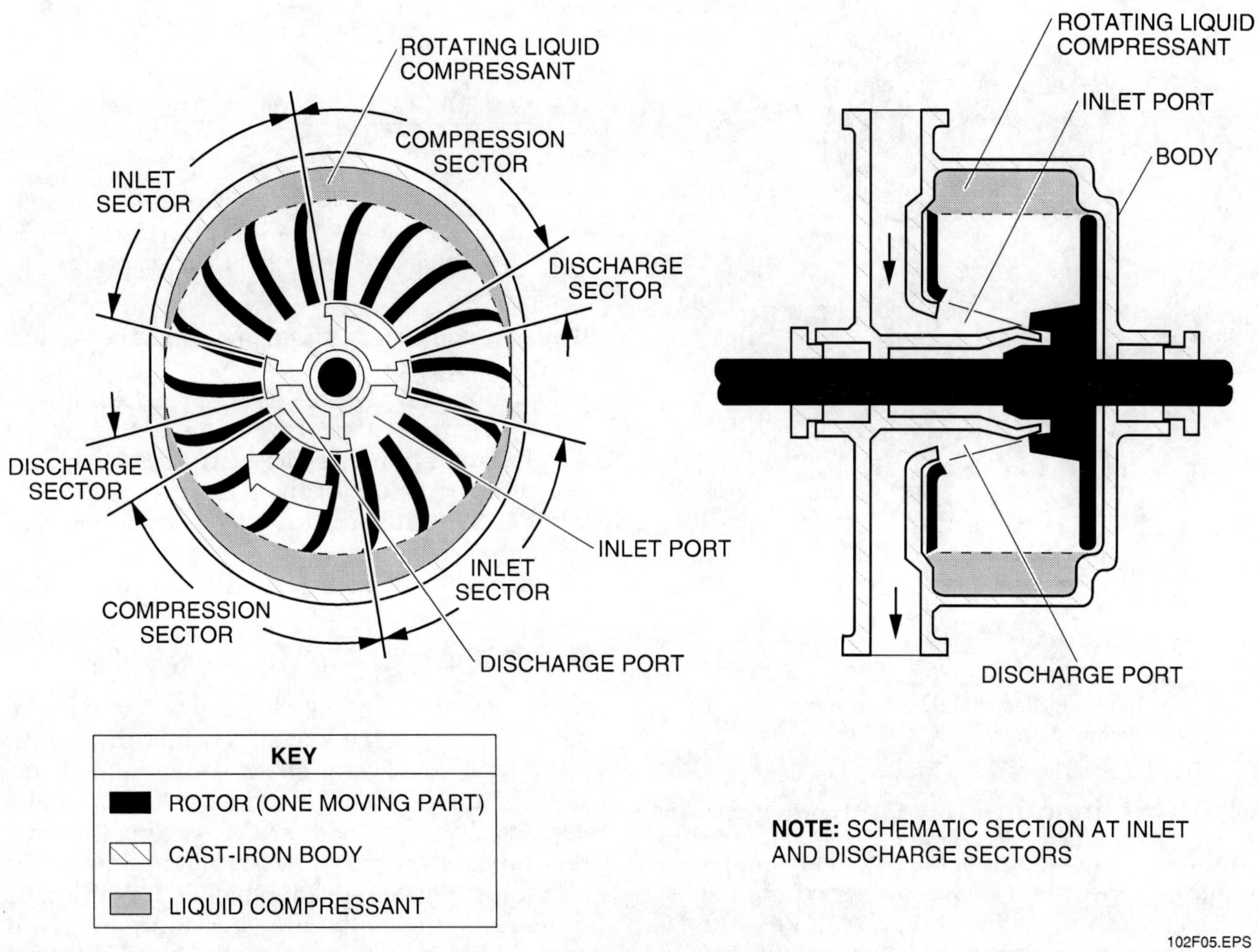

Figure 5 ◆ Typical liquid-piston rotary compressor.

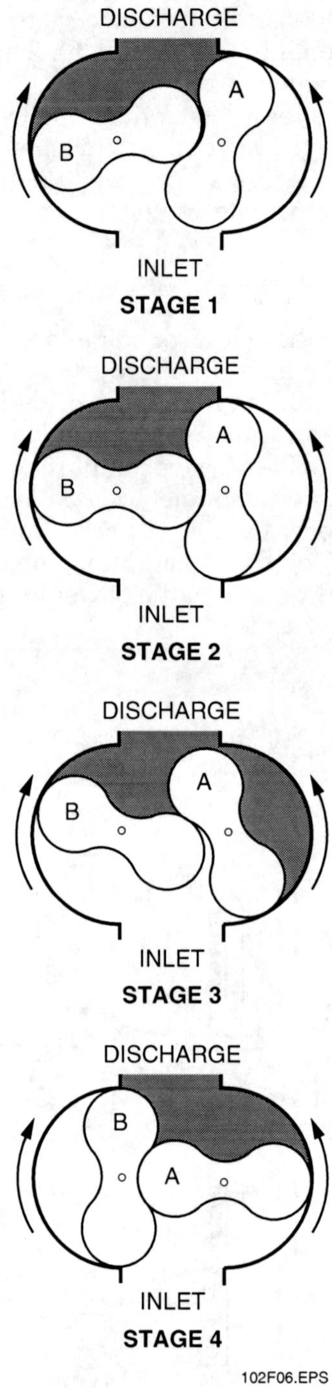

Figure 6 ◆ Twin-impeller, straight-lobe compressor operating cycle.

5.2.0 Continuous-Flow Compressors

Continuous-flow compressors contain a rapidly rotating element that boosts the speed of the gas as it passes through the compressor. Pressure is created partly in the rotating element and partly in the stationary **diffusers** or blades. There are two main types of continuous-flow compressors: ejector-type and dynamic.

Ejector-type compressors use a high-speed gas or steam jet to capture the flowing gas. The gas is then carried to a diffuser where speed is converted to pressure.

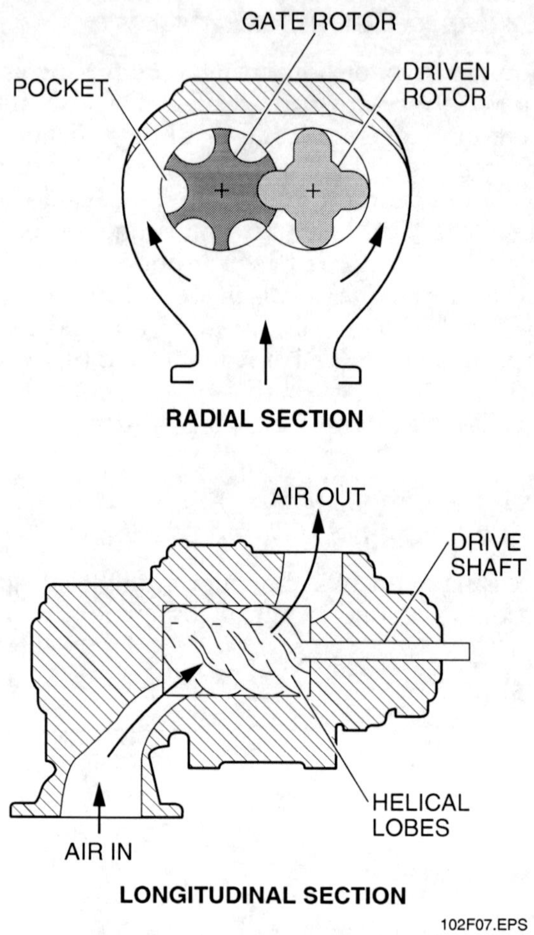

Figure 7 ◆ Sectional views of the helical compressor.

Dynamic compressors depend on the transfer of energy from a set of rotating blades to compress the gas. The rotor transfers this energy by changing the speed and pressure of the gas. The speed is converted into useful pressure energy as the gas moves through a stationary diffuser. There are two types of dynamic compressors: centrifugal and axial.

5.2.1 Centrifugal Compressors

In a cross-section view, the centrifugal compressor looks somewhat like a French horn. A single-stage centrifugal compressor is shown in *Figure 8*. This compressor has an impeller with radial, or backward-leaning, vanes. Gas is forced through the impeller by the rapidly rotating vanes. This rotation produces speed, which is converted into pressure partly in the impeller and partly in the radial diffuser. The impeller forces the gas through the radial diffuser and into the volute (spiral-shaped) collector. The pressurized gas circulates around the volute collector and flows away from the compressor, through the volute diffuser, and into the pipeline.

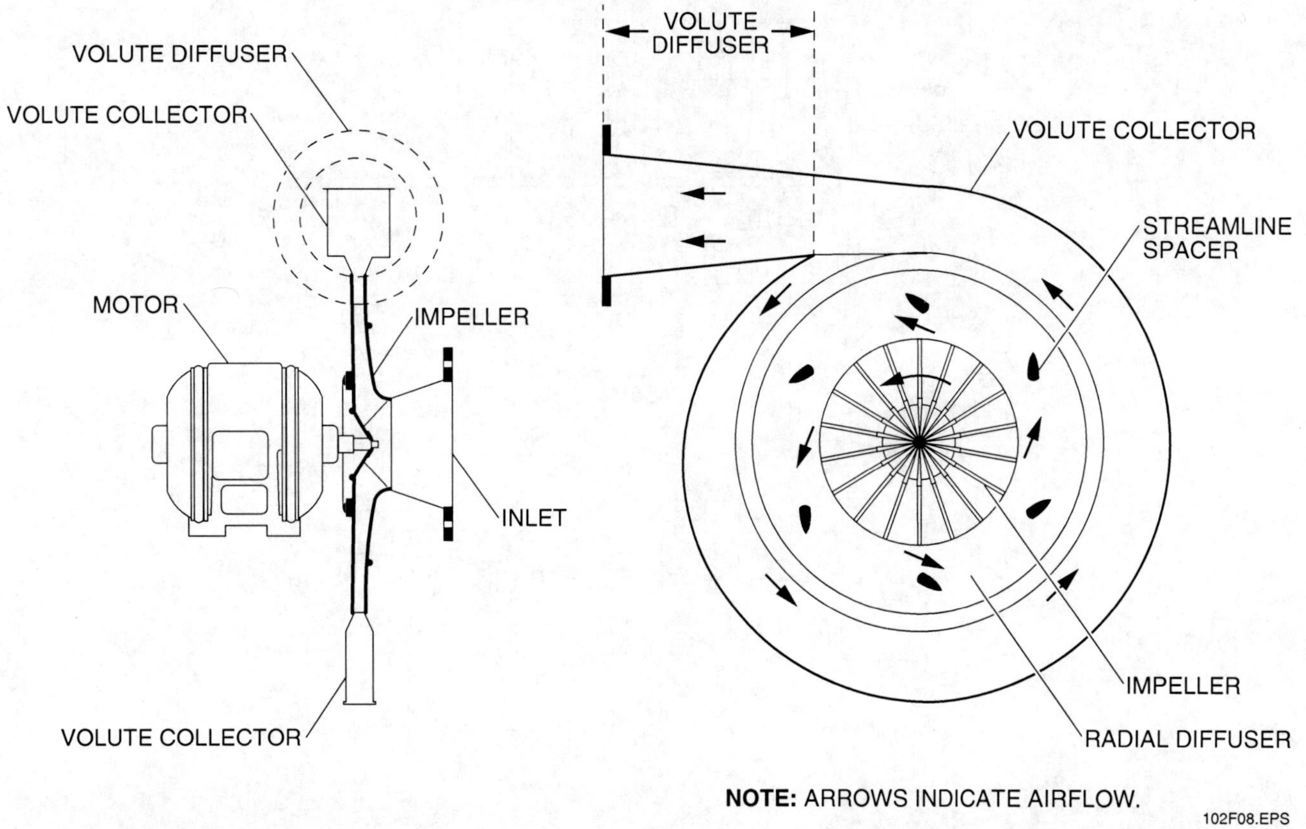

Figure 8 ◆ Single-stage centrifugal compressor.

5.2.2 Axial Compressors

An axial compressor is a large-capacity, high-speed machine (see *Figure 9*). Each stage in this compressor includes two rows of vanes: one rotating, one stationary. The rotating vanes, or rotors, increase gas speed. The stationary vanes, or stators, decrease gas speed. The speed produced by the rotor vanes is converted into pressure. This compressor does not create much swirling action, so gas flows in a straight line (also called axial direction).

5.3.0 Compressor Startup and Shutdown

Compressors in the gas pipeline must be started and stopped carefully. Gas compressor units are complicated and dangerous if not handled properly. You must carefully read and fully understand the manufacturer's startup and shutdown information.

WARNING!

Gas and air mixtures under pressure can explode. You can be severely injured or killed. Always follow the manufacturer's instructions for purging a compressor unit of any explosive mixture.

5.3.1 Startup

Pipeline personnel must perform a number of checks before starting up compressors. Most companies have a complete checklist that must be followed to ensure that the compressor is started safely. A wide variety of compressors are available, and different operating instructions go with each type. Therefore, you must always read and understand the manufacturer's technical and operations manual before starting a compressor. This manual includes proper installation instructions, startup checklists, and operating instructions. In general, startup checklists include the following operations:

- Ensure that the correct manufacturer's technical manuals are available at the work site for easy reference.
- Check the design limitations for the compressor, such as the operating pressure, discharge temperature, and recommended operating speeds.
- Review the design operating conditions.
- Ensure that all moving parts are working properly.
- Ensure that the gas high-discharge temperature shutdown device is working properly and is set at the correct percentage above the anticipated

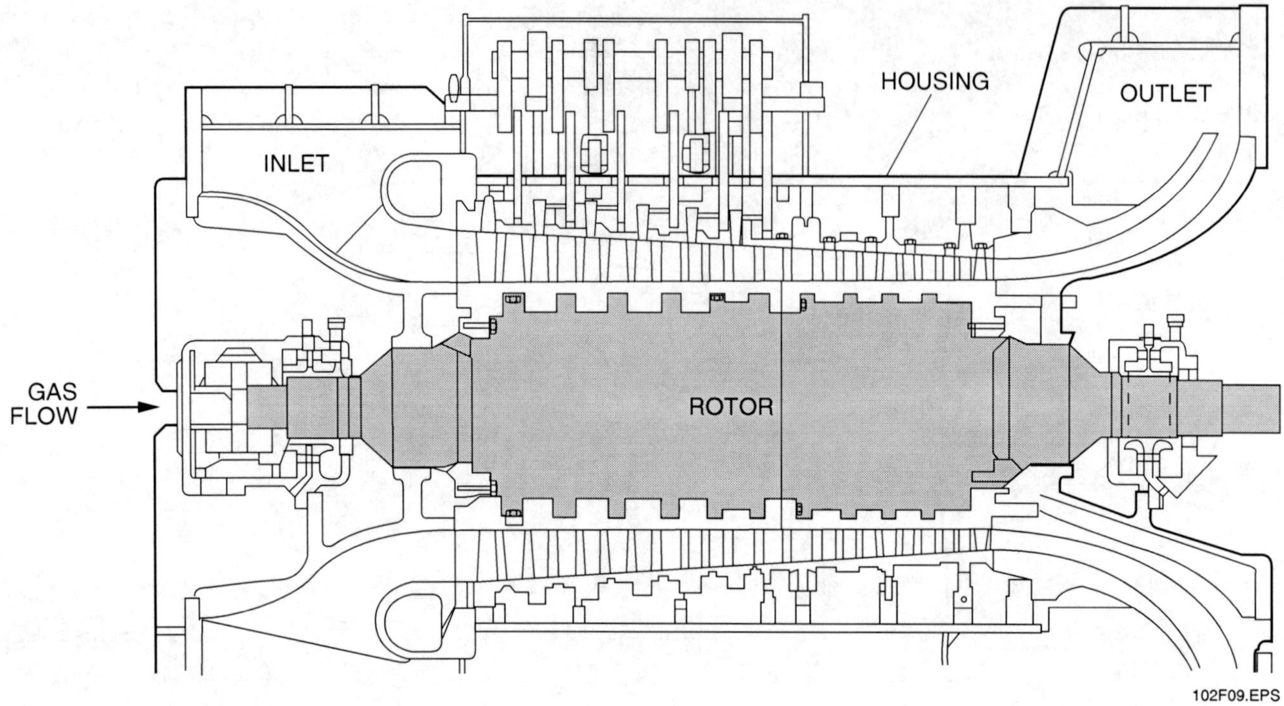

Figure 9 ◆ Axial compressor components.

temperature. This device monitors gas temperature at the discharge area and shuts down the compressor if the temperature exceeds a preset limit.

- Ensure that the strainer screen is installed **upstream** of compressor suction.
- Ensure that the compressor bearings and mechanical seal are properly lubricated.

> **CAUTION**
>
> Mechanical seals on compressors are designed to work with an oil film. Therefore, the seals must be lubricated before startup and be kept lubricated during operation. If the seals are allowed to dry out, they wear out more quickly.

- Ensure that the driver rotation matches the compressor rotation.
- Ensure that the piping and the compressor have been purged to remove all air and contaminants.

Many manufacturers also provide a log sheet on which details about maintenance can be recorded. Note that many manufacturers require that the log sheet be completed to maintain the warranty.

5.3.2 Remotely Controlled Shutdown Devices

Remotely controlled shutdown devices (RCSDs) may be operated by a signal sent from the pipeline's **control center.** A simplified example of how these devices work follows. A meter sends a pressure reading to a sensor in the control center. This reading is converted into information that is stored in a database. The system can then compare stored normal operating pressure against the current operating pressure. If the system detects a problem, a signal is sent to the RCSD to **shut in,** or isolate, part of the pipeline.

5.3.3 Emergency Shutdown Devices

By monitoring the line carefully and using pressure-relief valves, pipeline companies can protect the pipeline from too much pressure buildup and still maintain a safe flow of gas. In these situations, emergency shutdown devices (ESDs) may be used.

ESDs are heavy-duty valves designed to handle any pressure imbalance in the pipeline. Most are installed between a pressure-reducing regulator and **downstream** equipment. These devices allow pipeliners to quickly and safely shut in a section of the pipeline. ESDs shut down when a sensor detects that pressure has gone over a preset limit. An ESD may be designed to shut down in response to only low-pressure conditions, only high-pressure conditions, or both.

The activation of an ESD does not always signal an emergency condition. Generally, the following three conditions must be met:

- People may be harmed or killed.
- There may be extensive property damage.
- There may be extensive environmental damage.

Emergencies usually involve actions outside business as usual. For example, a buildup of pressure may be routinely handled by activating a relief valve. If the relief system fails, however, pressure may build up to emergency levels. Following are examples of conditions that may trigger an emergency situation:

- Pressure imbalance in the system
- Uncontrolled release
- Fire or explosion near or on the pipeline
- Bad weather conditions such as floods, tornadoes, hurricanes, and earthquakes
- Riots or violent protests on or near the pipeline
- Criminal or terrorist actions

In the event of an emergency, it is important to follow company policy for notifying local emergency response agencies. Many situations can be safely handled without causing undue alarm. Therefore, pipeline employees follow these general steps:

Step 1 Evaluate the situation to decide whether a true emergency exists.

Step 2 Notify the appropriate company personnel.

Step 3 Carry out established company procedures to protect people, property, and the environment.

Step 4 Notify the local response emergency agency if appropriate.

5.4.0 The Operation of Gas Compressors

It is important to understand the major parts of gas compressors. These include distance pieces, pistons, rings, and valves.

5.4.1 Distance Pieces

As gas is being compressed, some of it may get by the compressor's piston rings and packing. If the gas gets into the crankcase, a hazardous condition can result. Therefore, devices called distance pieces are installed to prevent excess gas, moisture, or lubrication from building up in the compressor area. A **stuffing box** keeps the gas from getting into the crankcase.

5.4.2 Compressor Pistons and Rings

Compressor pistons are made from cast steel, cast iron, aluminum, various alloys of machined steel, and steel alloys. Each pair of pistons should be weight balanced to reduce vibration. In single-stage compressors, achieving a complete balance is not possible. To offset the vibration, the counterweights on the crankshaft can be adjusted to improve performance, and a heavier crankcase and frame can be added.

In every reciprocating compressor, a set of rings seals the gas. The rings also keep the piston from rubbing against the cylinder wall and wearing out too fast. In piston-type compressors, lubrication must be pumped into the compression cylinder to lubricate the piston rings. Providing some of the same lubrication to the packing extends its life.

5.4.3 Compressor Valves

Suction and discharge valves in the head of the compressor control the gas intake and discharge of the compressor cylinder. The valve area (total open area when the valve is open) affects the compressor's efficiency. The proper valve design minimizes pressure drop across the valve and helps the unit to operate efficiently. Selecting valves carefully also increases the efficiency of the compressor. Valves with a lower spring tension open more easily and operate more efficiently.

6.0.0 ◆ COMPRESSED-AIR TREATMENT

Air entering a compressor is often hot and contaminated, which reduces compressor efficiency. Therefore, the air must be treated. The following devices are used to remove contaminants and regulate temperature:

- Intake filters
- Intercoolers
- Aftercoolers
- Separators
- Chemical dryers
- Receivers
- In-line filters
- Pressure regulators
- Lubricators
- Filter-regulator-lubricators

6.1.0 Intake Filters

Compressors work best when the air flowing into them is cool, dry, and clean. However, air contains dirt and dust, which, if allowed to get into the compressor, cause it to overheat and lose efficiency. Intake filters, which must be cleaned or replaced on a regular schedule, solve this problem. Follow the manufacturer's instructions and recommendations for cleaning and replacing filters. The two main types of intake filters are dry and wet.

The dry intake filter works a lot like an air filter on a car. As air flows through the filter, it traps dirt and dust (see *Figure 10*).

In the wet intake filter, air passes over oil before it enters the filter (see *Figure 11*). The oil traps dirt and dust and helps remove moisture from the air before it enters the compressor. Wet intake filters are rarely used on modern equipment, but some may still be found on older equipment.

> **NOTE**
> Filters are rated by the size of the openings in the material. This rating is based on the size of the largest particle that can pass through the filter. Filtration is measured in microns. One micron is 39 millionths of an inch (0.0000039). The size of filter to use for a particular application depends on the micron size and the flow volume.

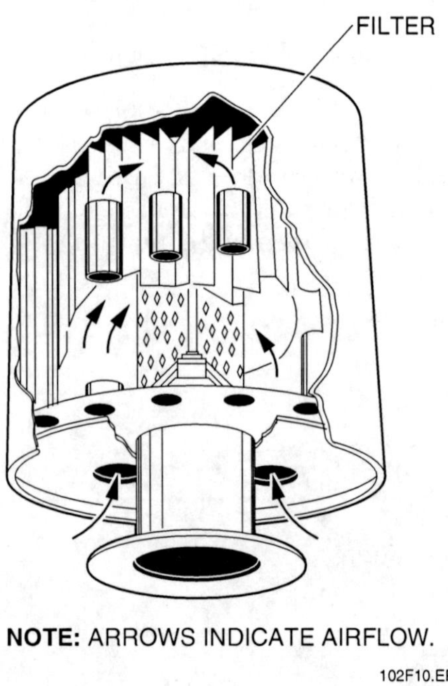

Figure 10 ◆ Dry intake filter.

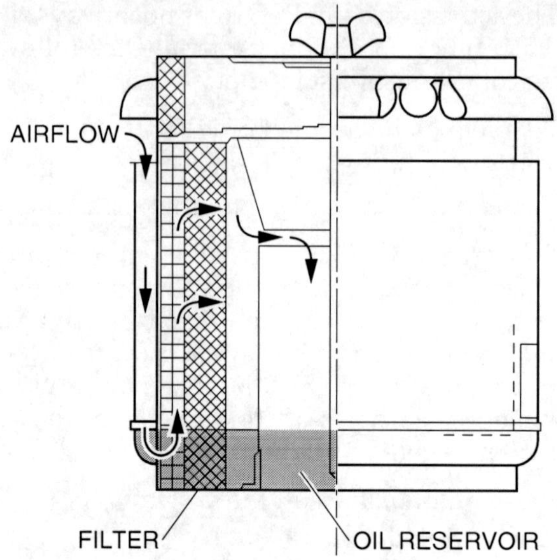

Figure 11 ◆ Wet intake filter.

6.2.0 Intercoolers

Compressing a gas produces heat, and heat reduces compressor efficiency. To solve this problem, the gas is cooled between compression stages with intercoolers. The two types of intercoolers used in multistage compressors are air and liquid.

The air intercooler has finned tubing that runs between the stages of the compressor (see *Figure 12*). As the gas hits the tubing, the heat is scattered, and the gas cools off.

The liquid intercooler uses water to cool the air. The water circulates through tubing inside a tank that contains the compressed gas.

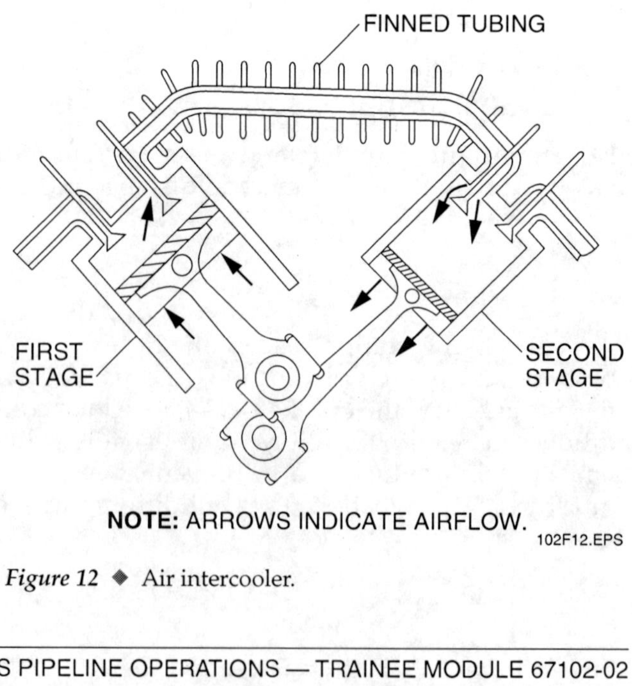

Figure 12 ◆ Air intercooler.

6.3.0 Aftercoolers

Aftercoolers lower the discharge temperature of compressed gas and remove oil vapor and water vapor. The main types of aftercoolers are air-cooled and water-cooled types and refrigeration dryers.

In the air-cooled aftercooler, compressed gas passes over finned tubing on its way to the air receiver. A fan-type flywheel draws the gas over the tubes, cooling it and condensing the moisture. The moisture passes on to a receiver and is drained either manually or by an automatic drain trap.

In the water-cooled aftercooler, a steel shell encases copper tubing and cooling fins (see *Figure 13*). Headers on each end of the aftercooler provide piping connections for water, air, and condensate piping.

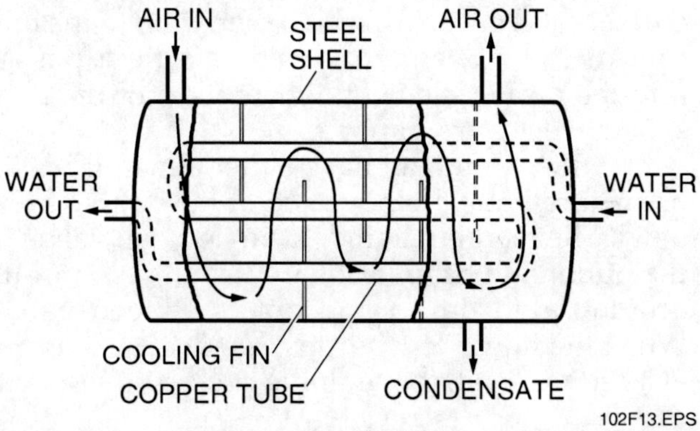

Figure 13 ◆ Water-cooled aftercooler.

In the refrigeration dryer, humidity is lowered to dry the gas. The refrigeration system captures moisture from the compressed gas as it passes through the heat exchanger and discharges it through a drain (see *Figure 14*).

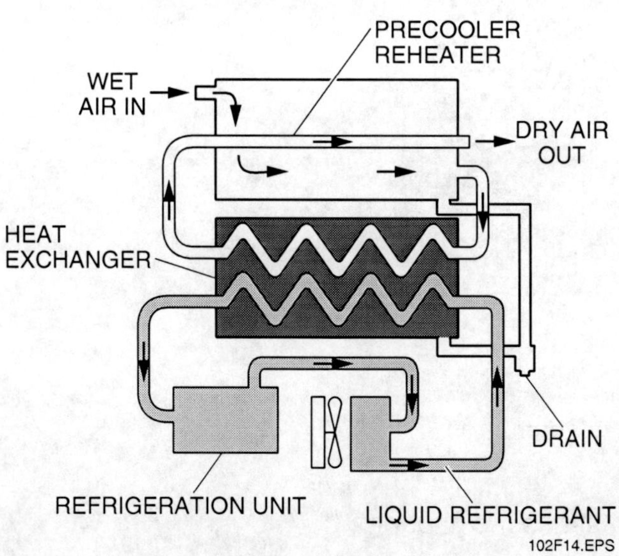

Figure 14 ◆ Refrigeration dryer.

6.4.0 Separators

Natural gas may contain other hydrocarbons and water that condense during transmission. Separators, which contain scrubbers or filters, remove these elements from the gas at the compressor station (see *Figure 15*). The gas usually enters the separator before entering the compressor station.

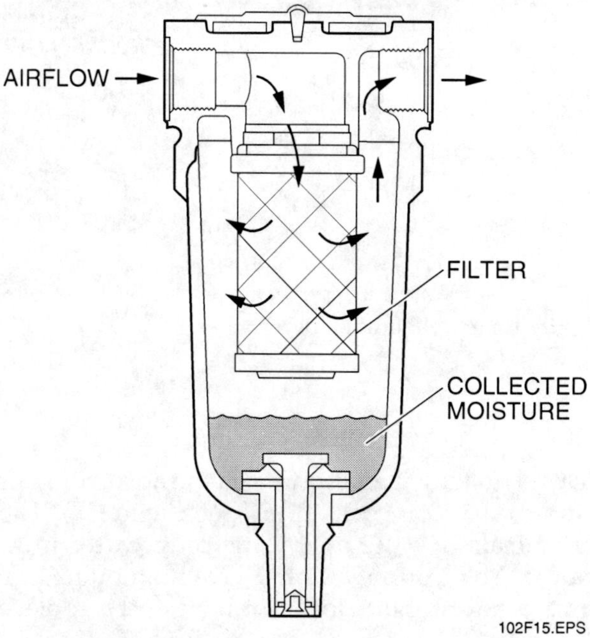

Figure 15 ◆ Separator.

6.5.0 Chemical Dryers

Chemical dryers clean and dry the air. They generally are placed in the pipeline following the separator. The two main types of chemical dryers are the deliquescent and the desiccant. The deliquescent dryer is an **absorption** dryer; the desiccant dryer is an **adsorption** dryer.

The deliquescent dryer uses a chemical to absorb water from the system (see *Figure 16*). As it absorbs the water, this chemical slowly melts and dissolves; therefore, it must be replaced on a regular basis. Some absorption dryers use two-stage systems with heaters to help the water evaporate. The two-stage system allows one side of the dryer to dry out while the other side is in use. Wet air flows into the dryer through a four-way valve that controls which side of the dryer is used. Heating elements dry the air, which then flows back through the four-way valve into the line. At the same time, check valves control the flow of condensate that comes out of the bottom of the dryer.

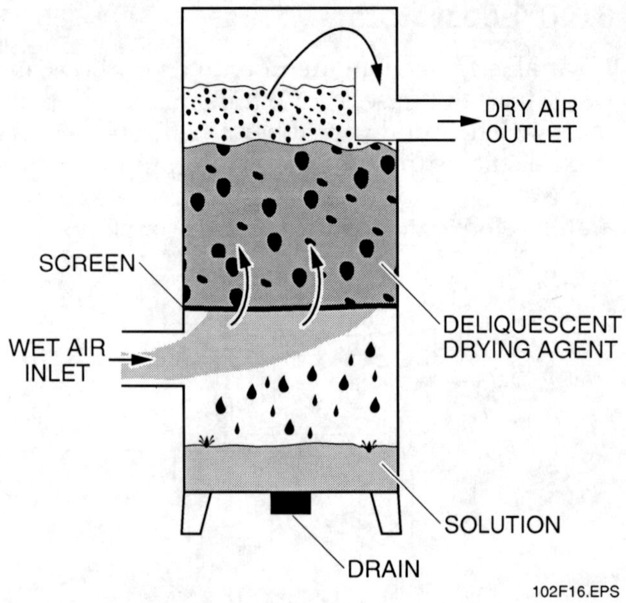

Figure 16 ◆ Deliquescent dryer.

The desiccant dryer uses chemicals, such as silica gel or activated alumina, in the adsorption process to dry compressed air (see *Figure 17*). These chemicals dry the air by trapping water in tiny pores. Adsorption systems have no moving parts, and the desiccant does not have to be replaced. Generally, however, an adsorption system costs more to run than an absorption system does.

6.6.0 Receivers

A receiver is a tank-like structure that stores compressed air (see *Figure 18*). It removes heat from the compressed air and collects moisture. The receiver acts as a buffer and allows air to flow smoothly and quietly from the compressor to the pneumatic system.

6.7.0 In-Line Filters

In-line filters (see *Figure 19*) remove most solid and liquid particles from compressed air. As air enters the filter unit, louvers force it into a swirling pattern. Centrifugal force separates the coarse, solid particles, which then drop to the bottom of the filter's bowl. The air then passes through a filter element, where most of the finer solid particles are trapped.

The filter unit may have either a manual drain or an automatic drain located at the bottom of its bowl. The manual drain is a single small valve. In the automatic drain, a float rises in response to accumulated liquid and opens a bleed valve. When the liquid level drops, the float also drops; at a certain level, the draining process is stopped.

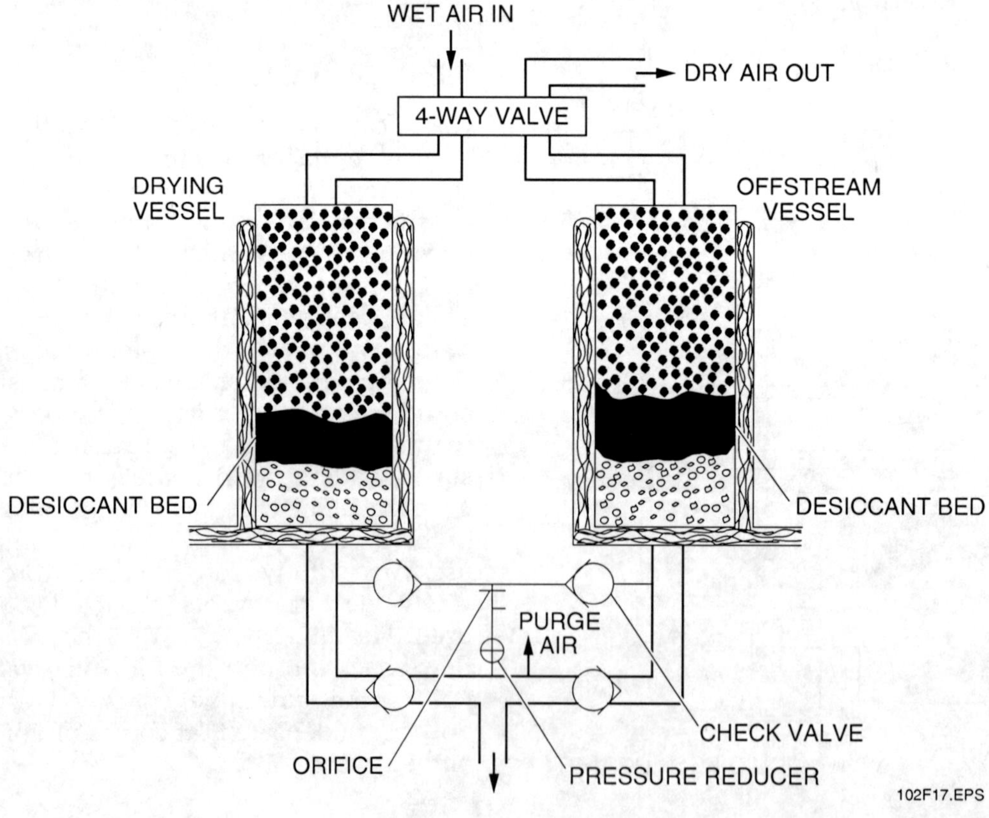

Figure 17 ◆ Desiccant dryer.

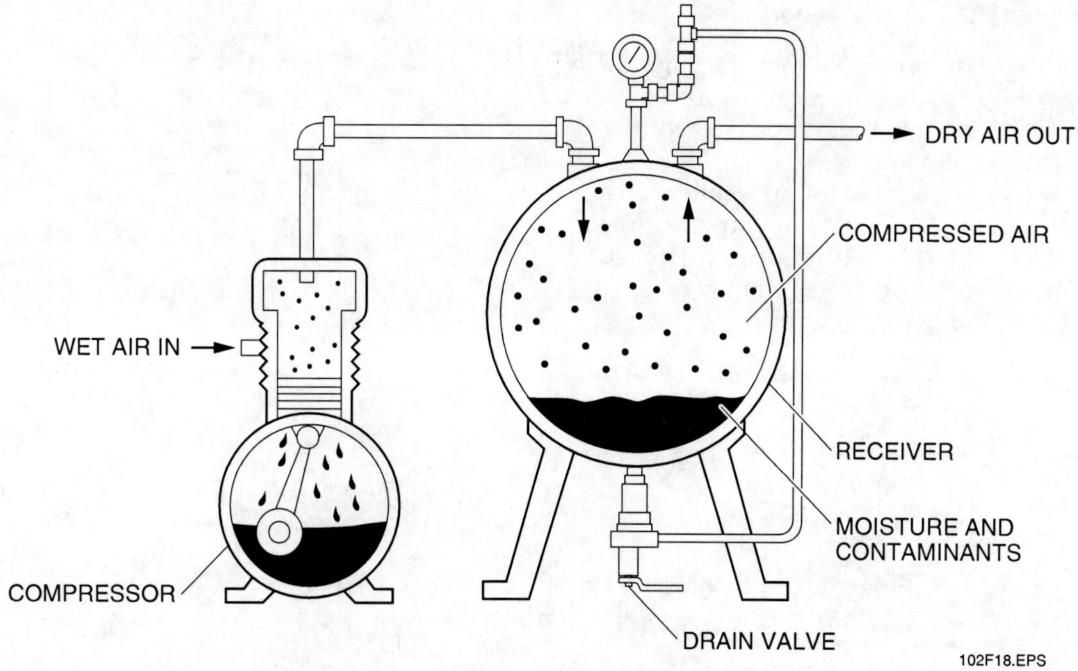

Figure 18 ◆ Receiver.

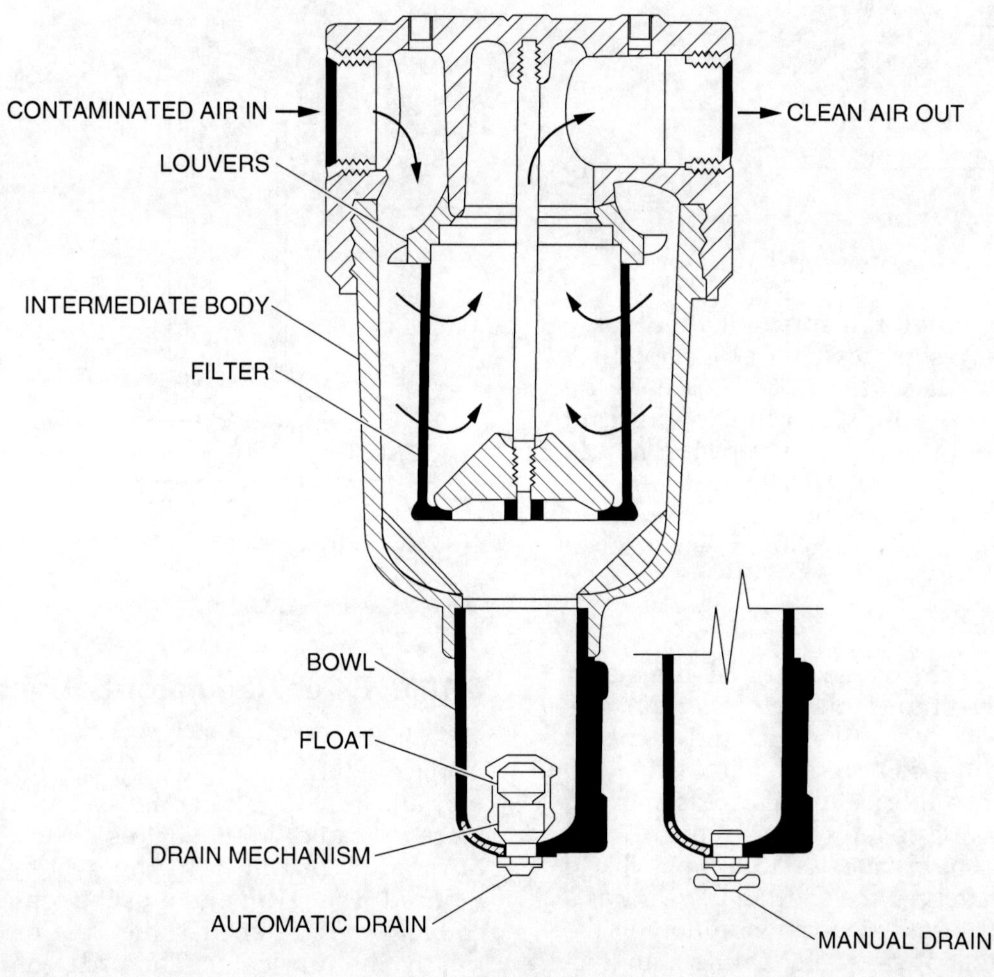

Figure 19 ◆ In-line filter.

6.8.0 Pressure Regulators

A pressure regulator controls the operation of a valve that either opens or closes to control pressure in the line. The two main types of regulators are diaphragm and piston.

In a diaphragm regulator, pressure either increases or decreases against a thin diaphragm (see *Figure 20*). As the downstream pressure increases, the diaphragm closes the valve. As the pressure drops, the diaphragm opens the valve.

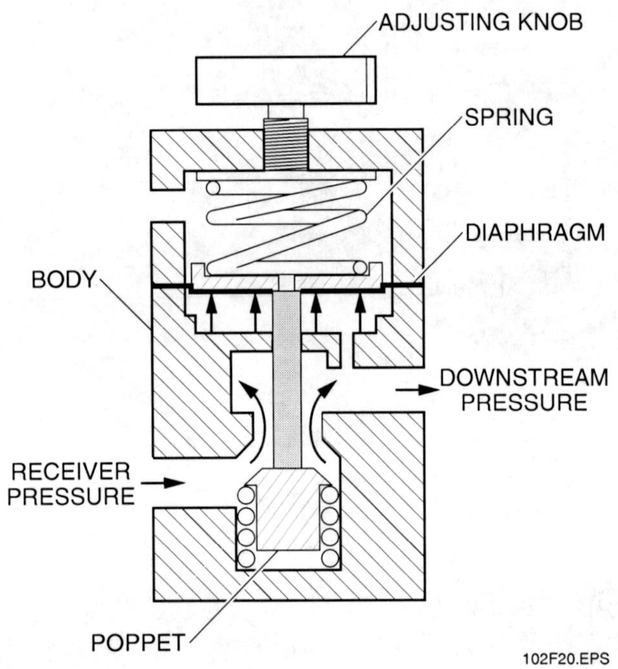

Figure 20 ◆ Diaphragm regulator.

In a piston regulator, a spring-loaded valve responds to changes in pressure (see *Figure 21*). A pressure increase is sensed in an area near the underside of the piston. When the downstream pressure nears a preset level, the piston moves upward and pulls the **poppet** into its **seat.** Once the poppet is seated, a constant source of compressed air is made available to an **actuator** downstream.

6.9.0 Lubricators

Lubricators add a small amount of oil into the air flowing through compressors (see *Figure 22*). The air breaks the oil into a fine mist and carries it downstream to the compressor. Oil is added to compressed air to lubricate seals, keep moving parts from sticking together, and reduce wear. Lubricators must be installed as close as possible to the components to keep the oil from traveling as a liquid. Incorrect installation can stop flow in the system. You must always refer to the manufacturer's manual to determine the amount and type of lubrication to use.

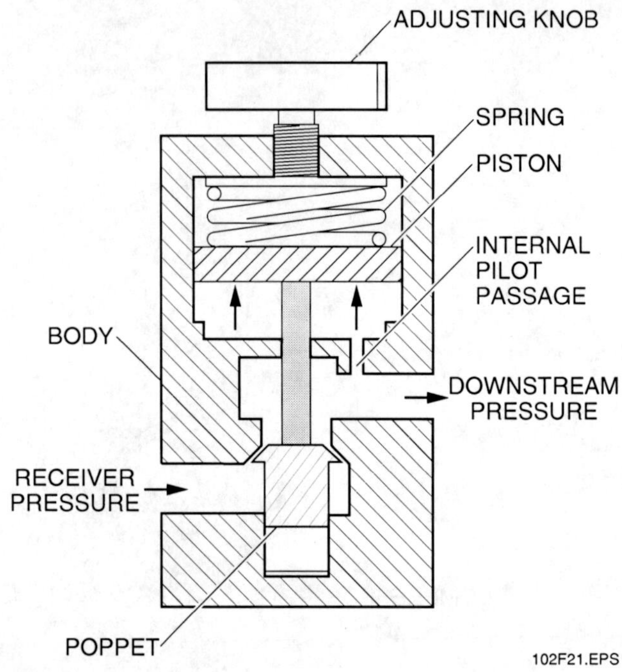

Figure 21 ◆ Piston regulator.

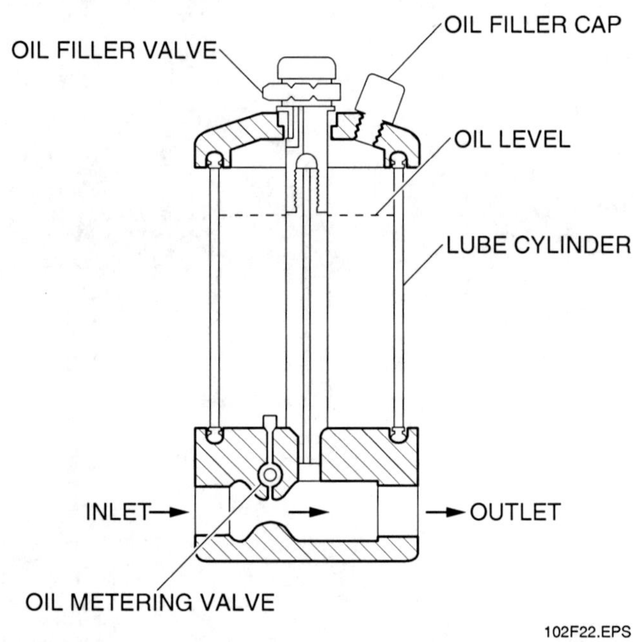

Figure 22 ◆ Lubricator.

6.10.0 Filter-Regulator-Lubricators

The filter-regulator-lubricator (F-R-L) unit is used in place of three separate units (see *Figure 23*). This unit allows you to take the system apart in one place instead of in three; thus, the chance of contaminants getting into the system is greatly reduced. The F-R-L unit must be connected in the correct order: filter first, then the regulator, and finally the lubricator. The order of connection must be followed to prevent damage resulting from contamination of the pneumatic system.

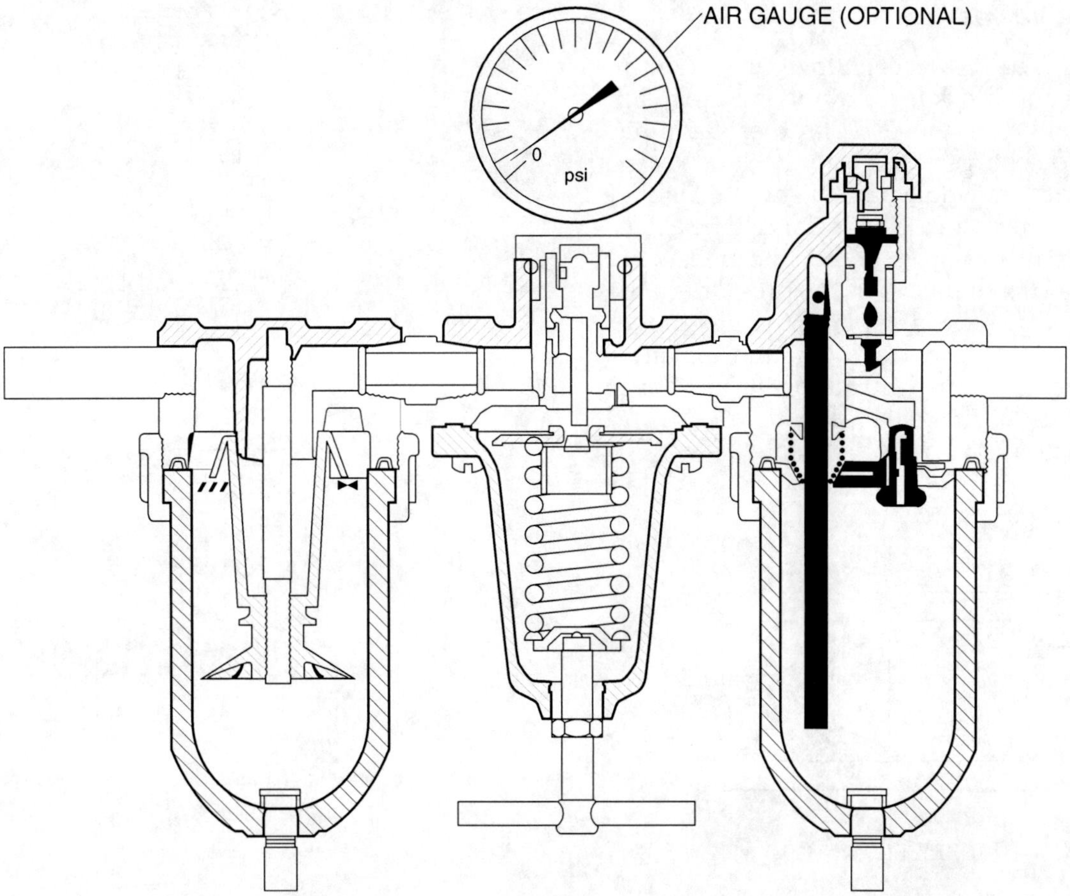

Figure 23 ◆ F-R-L assembly.

7.0.0 ◆ PNEUMATIC SYSTEM COMPONENTS

Pneumatic system components are designed to control flow, operate system equipment, and muffle noise. The main pneumatic system components are as follows:

- Valves
- Actuators
- Mufflers

7.1.0 Valves

Valves are grouped by how they control airflow. Valves may direct, check, block, or restrict airflow. They may also control or relieve pressure in the line. Following are the five most common types of valves:

- Directional-control valves
- Check valves
- Flow-control valves
- Pressure-control valves
- Pressure-relief valves

NOTE

Each valve that might be needed during an emergency must be inspected and operated on a regular schedule. Government regulations state that these inspections must take place at intervals not greater than 15 months, but at least once each calendar year (*49 CFR 192.745*).

7.1.1 Directional-Control Valves

Directional-control valves allow or prevent airflow through the pipeline and come in a wide range of sizes and types. This type of valve may be controlled manually, mechanically, or with a pilot device. Manual controls include push buttons, switches, push-pull levers, or foot pedals. In valves that are mechanically controlled, an electrical signal trips a **solenoid,** which causes another solenoid to operate. A push button, limit switch, or sequencing relay may activate the electrical signal. Pilot devices are activated by a pressure change in the circuit that makes the valve open or close.

7.1.2 Check Valves

Check valves (see *Figure 24*) allow air to flow in one direction and block it from flowing upstream or back through the line. This type of valve has suction and discharge ports at either end of the valve body and a movable part, or seal, that is activated by spring pressure. The seal might be a flapper, plunger, ball, or poppet. When the pressure at the suction port is higher than the pressure of the spring that holds the seal in place, the seal moves off its seat. Air then flows freely through the valve. When air tries to return through the discharge port, the air pushes the seal back onto its seat, and airflow through the valve is blocked.

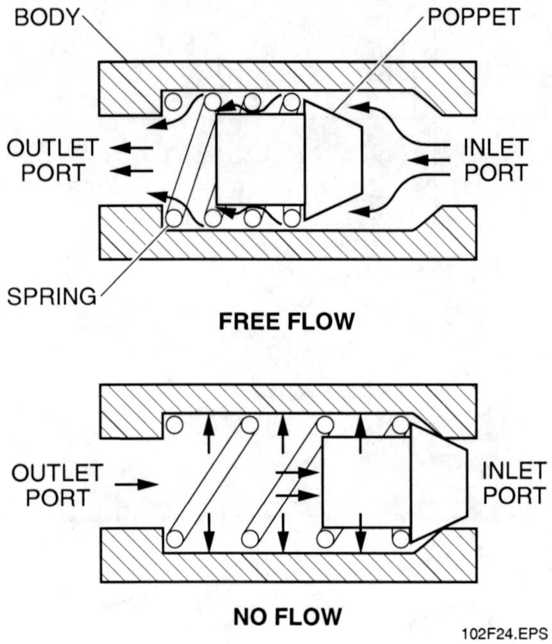

Figure 24 ◆ Check valve.

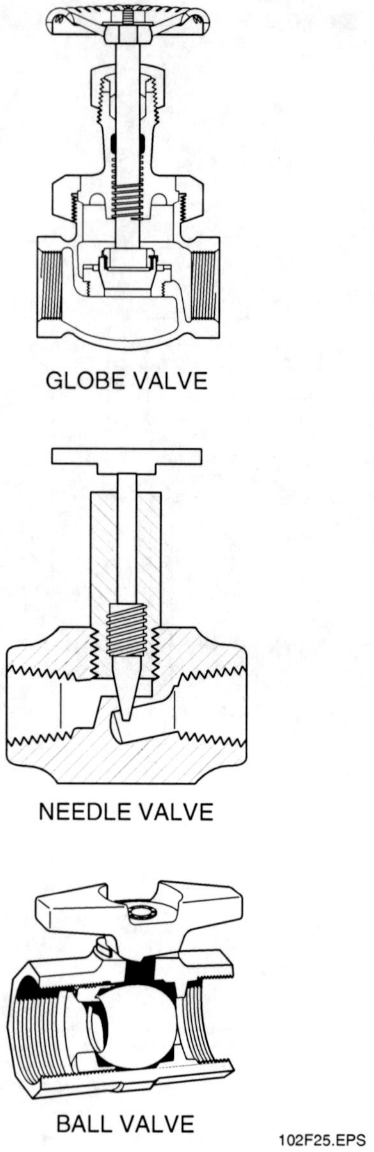

Figure 25 ◆ Flow-control valves.

7.1.3 Flow-Control Valves

Flow-control valves restrict airflow. Controlling how much air flows through the valve helps control the operating speed of the actuator. A flow-control valve might be fixed or adjustable. Three common flow-control valves are globe, needle, and ball valves (see *Figure 25*).

Air does not flow in a straight line through a globe valve. It must turn 90 degrees and pass through the orifice, or opening, that is the seat of the globe. The orifice size is adjustable and determines the rate of flow.

A common type of flow-control valve in a pneumatic system is a combination needle valve and bypass check valve. Air flowing through the needle valve must turn 90 degrees to pass through the orifice. The orifice is the seat for a cone-shaped rod, and the position of the rod in relation to the seat determines the orifice size. Fine threads on the valve stem allow gradual changes in the orifice size. The bypass check valve allows reverse airflow in the system.

A ball valve allows air to flow straight through its center. A handle is used to change the orifice size. As the handle is turned, a ball with a cross-drilled hole is positioned across the airflow path. The size of the hole determines the flow rate.

7.1.4 Pressure-Control Valves

Compressed air held in a compressed-air storage tank must be portioned out to the actuators downstream. Pressure-control valves control the pressure at two points in the system: after the compressor and after the air receiver tank.

Pressure-sensing devices automatically control the amount of pressure produced by the compressor. To manage the compressor's operating range, the system signals the compressor to turn on or off. If this system fails, a safety-relief valve vents some pressure to the atmosphere. This venting process allows the compressor to return to a safe operating level.

7.1.5 Pressure-Relief Valves

Pressure-relief valves (see *Figure 26*) are sometimes called unload or pop-off valves. These valves are safety devices that protect the line from a buildup of too much pressure. Pressure-relief valves should be installed close to the compressor or the system part for which they relieve pressure. The valve is preset to a specific pressure point. Pressure above this point overcomes a low-pressure spring that keeps the valve closed. When the valve opens, it vents air or gas to the atmosphere and reduces the pressure buildup. When the pressure decreases to the preset point, the spring closes the valve.

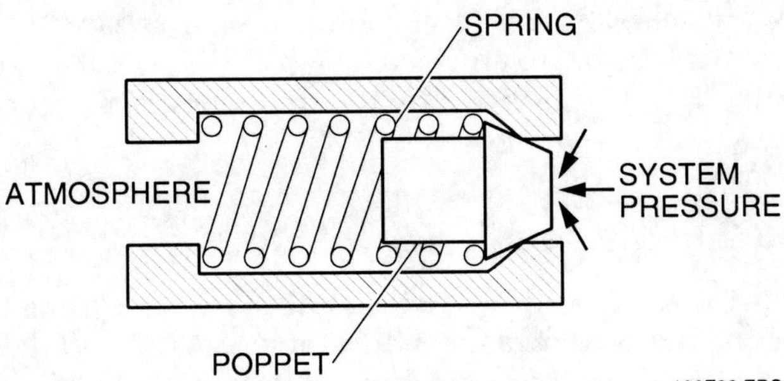

Figure 26 ◆ Pressure-relief valve.

7.2.0 Actuators

Actuators are mechanical devices that move or control the operating parts of a pneumatic system. Common examples of actuators include an air-driven dentist's drill and a pneumatic lift at a service station. Two types of actuators in pneumatic systems are pneumatic cylinders and pneumatic motors.

7.2.1 Pneumatic Cylinders

Pneumatic cylinders channel air pressure in a straight line. Air pressure in the cylinder pushes a piston forward to an area where there is less pressure. Pneumatic cylinders are found in jackhammers, drill presses, and other impact-type tools.

Pneumatic cylinders fall into two categories: single acting and double acting. The single-acting cylinder applies force to move the piston in one direction. A weight or spring pushes the piston back again. The double-acting cylinder applies force in both directions.

7.2.2 Pneumatic Motors

Pneumatic motors convert the force of compressed air into mechanical energy. The vane motor (see *Figure 27*) is the most common type of pneumatic motor. This motor is made up of vanes, a rotor, a shaft, a ring, and suction and discharge ports. Compressed air enters the suction port and pushes against the vanes, which causes the motor to produce an output **torque.** As air pressure against the vane increases, rotation also increases.

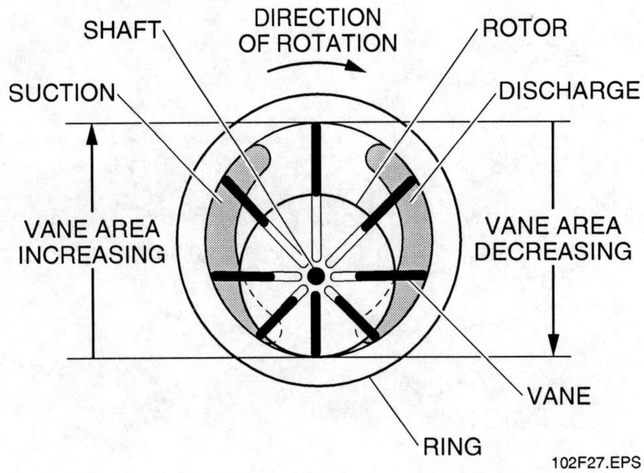

Figure 27 ◆ Vane motor.

7.3.0 Mufflers

When pressure is released from a pneumatic system, it makes a lot of noise. Much like mufflers on cars, mufflers are used on air systems to reduce noise. However, mufflers also restrict airflow, which can reduce system efficiency. Only mufflers that are compatible with the system should be used. If the muffler is not compatible, backlash is created in the system, which could damage the compressor. You must follow the manufacturer's technical specifications to ensure the right match between the muffler and the system.

CAUTION

Mufflers may not reduce noise levels enough to protect hearing. In these cases, earplugs must be worn. Follow company safety rules regarding the use of proper hearing protection.

Review Questions

Sections 5.0.0–7.0.0

1. A compressor increases the pressure of gas by _____.
   a. reducing its volume
   b. increasing its volume
   c. heating it up
   d. cooling it down

2. The two main types of compressors are _____.
   a. helical and axial
   b. reciprocating and sliding vane
   c. positive displacement and continuous flow
   d. negative displacement and intermittent flow

3. In a liquid intercooler, the air is cooled with _____.
   a. oil
   b. silica gel
   c. nitrogen
   d. water

4. _____ lower the discharge temperature of compressed gas and remove oil vapor and water vapor.
   a. Intercoolers
   b. Aftercoolers
   c. Separators
   d. Receivers

5. _____ valves allow air to flow in one direction and block it from flowing upstream, or back through the line.
   a. Gate
   b. Block
   c. Check
   d. Needle

8.0.0 ◆ METERS

Three basic types of meters are used on the gas pipeline: turbine, ultrasonic flow, and orifice. Meters are the cash registers of the pipeline industry. They record how much gas goes from one point to another, allowing the pipeline company to track deliveries and bill customers. Therefore, meters must be calibrated before use and checked on a regular basis to ensure accurate measurement. The three main costs associated with providing gas to customers are as follows:

- *Transmission cost*—The cost to move the gas from its source, through the pipeline, to a local gas distributor
- *Distribution cost*—The cost to move the gas from the local gas distributor to individual customers
- *Commodity cost*—The cost of the gas itself

8.1.0 Turbine Meters

Gas turbine meters are used for the **custody transfer** of natural gas (see *Figure 28*). They are also used in distribution and supply stations. These meters are designed to operate with high-energy efficiency at low gas speed. A turbine rotor guides the gas flow through the meter to an exit channel, which is designed to minimize pressure loss.

102F28.EPS

Figure 28 ◆ Gas turbine meter.

Turbine meters measure speed and convert it to volume. Rotors in the meter have a sensor that sends a signal to an electronic counter. The total of the signals is divided by the frequency per unit of volume.

8.2.0 Ultrasonic Flow Meters

An ultrasonic flow meter is also called a transit-time meter. It measures the amount of time it takes for an ultrasonic sound wave to transit, or move, from one sensor to another. The transit time is measured in both the upstream and downstream directions. Ultrasonic flow meters come in two basic types: single path and multipath (see *Figure 29*).

102F29.EPS

Figure 29 ◆ Multipath ultrasonic flow meter.

In the single-path meter, the transit time is measured with a single set of sensors. The transit time (stated in feet per second) is calculated by dividing distance by time.

A multipath ultrasonic meter works exactly like a single-path meter. However, the multipath meter uses several pairs of sensors to send and receive signals. Because more information is collected, multipath readings tend to be more accurate than single-path readings.

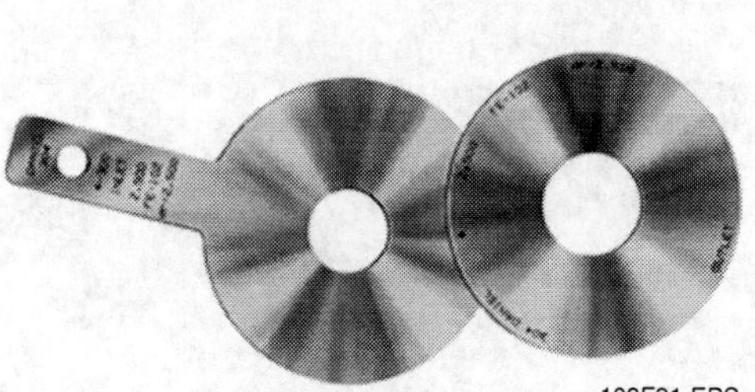

102F31.EPS

Figure 31 ◆ Orifice plates.

8.3.0 Orifice Meters

Orifice meters (see *Figure 30*) restrict the flow of gas, thus producing a difference in pressure. That pressure difference is measured to calculate actual flow. The orifice meter comes in a wide variety of sizes and types. Some models are designed for repair in the field, which makes them highly efficient for pipeline use.

102F30.EPS

Figure 30 ◆ Orifice meter.

Orifice plates (see *Figure 31*) allow precise flow of natural gas through the meter. These plates are set in the body of the meter housing, and you can easily remove them for maintenance. The plates of some meters may be removed without stopping the flow of gas. This feature saves time and money.

8.4.0 Meter Protection

Meters are sensitive units that must be kept in good working condition. Meters are equipped with protection and control equipment, including alarms, flow controllers, pressure controllers, strainers, and air ventilation equipment.

Flow controllers manage the flow of gas through the meter, which helps to protect the meter from a buildup of pressure. Flow controllers also prevent gas from flowing in the wrong direction, or back through the meter. Pressure gauges are designed to detect a change in pressure caused by a clog in the strainer. A sensor on the pressure gauge sends an alarm to the control center if the strainer clogs. Air ventilation equipment is installed upstream of meters to vent excess air. If excess air starts to build up in a meter, an alarm is sent to the control center.

9.0.0 ◆ FLOW AND TEMPERATURE CONTROLS

It is important to monitor and regulate gas flow and temperature as the gas moves through the pipeline. Faster-moving gas heats up more quickly and must be directed to cooling towers located between compressors before being returned to the main pipeline. In addition, gas may be heated for certain processes.

9.1.0 Flow Control

Flow controllers (see *Figure 32*) monitor and measure flow through the pipeline around the clock. Some flow controllers are stand-alone units. Others are networked to a data collection system. The information collected from the flow controller is processed and stored in the unit. Personnel can read this information in the field and send it by telephone, radio, or satellite transmission to other locations. Flow controllers are installed at various places along the line, including compressor stations, custody transfer stations, and gate stations. These controllers may use turbine meters, ultrasonic meters, or orifice plates to monitor the flow of gas.

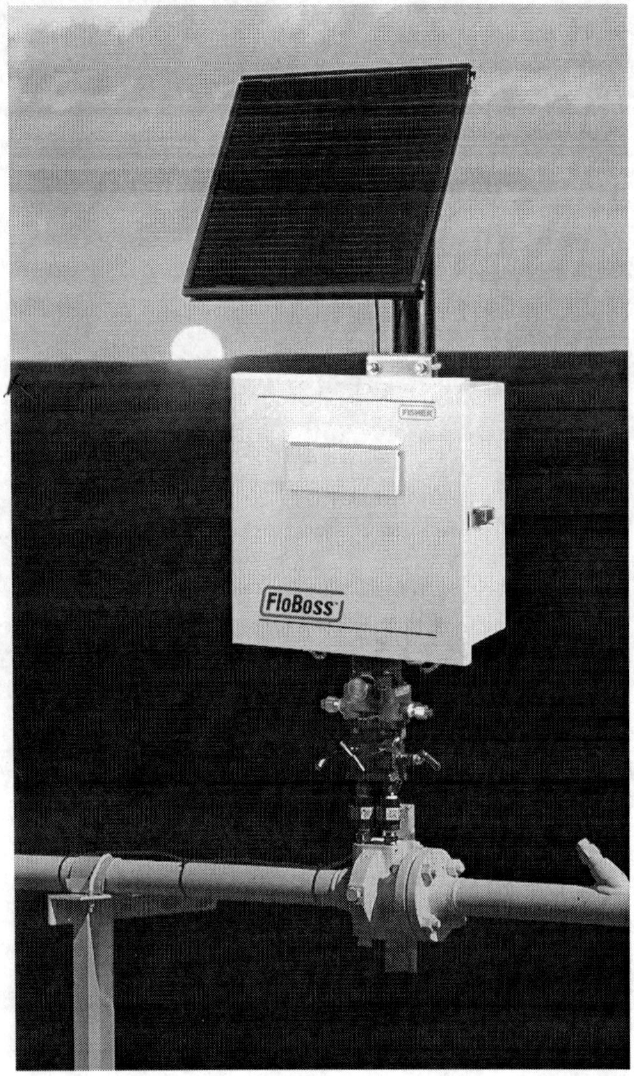

102F32.EPS

Figure 32 ◆ Gas flow controller.

9.2.0 Temperature Control

As you have learned, compression causes gas to become hotter. In the transmission line, gas can be cooled to acceptable levels in several ways. On land, the gas moves through cooling stations where water or air is used to bring the temperature down to acceptable levels. The gas also may be cooled when the cooler ground temperature or cooler water temperature surrounding the pipeline draws off heat.

Gas may have to be heated for process control, such as to operate a valve, compressor, or other line equipment. The temperature is increased on the upstream side to power the equipment and then decreased on the downstream side. A programmable logic controller in the control center is used to control the process and the compressors.

Review Questions

Sections 8.0.0–10.0.0

1. The three basic types of meters used on gas pipelines are _____.

a. turbine, ultrasonic flow, and orifice
b. intercoolers, aftercoolers, and separators
c. master, reciprocating, and sliding vane
d. positive displacement, continuous flow, and axial

2. The three main costs associated with providing gas to customers are _____.

a. research, development, and compliance
b. transmission, distribution, and commodity
c. research, transportation, and development
d. transmission, transportation, and maintenance

3. Turbine meters measure speed and convert it to _____.

a. energy
b. heat
c. volume
d. temperature readings

4. Orifice meters _____ the flow of gas.

a. redirect
b. restrict
c. measure
d. monitor

5. Flow controllers may use _____ to monitor the flow of gas.

a. turbine meters, ultrasonic meters, or orifice plates
b. pressure gauges, check valves, or gate valves
c. check valves, gate valves, or needle valves
d. compressor pressure or flow volume

10.0.0 ◆ ODORIZATION EQUIPMENT

Natural gas must be odorized in distribution lines so that if a leak occurs, anyone with a normal sense of smell can detect it. Generally, **odorants** are not required in transmission lines, offshore lines, or lines that run through sparsely populated areas. Odorant, usually a chemical called mercaptan, is added to natural gas, which is odorless, to make it smell like rotten eggs.

There are three main types of odorization equipment: bypass, wick, and electronic. Each diverts the flow of gas across a sensor that triggers the flow of odorant into the gas stream. Personnel use monitors inserted into the gas stream to monitor the gas for correct odorant levels.

Generally, it is the responsibility of the local distribution company to add odorant to the distribution line. Computerized units are available that allow responsible personnel to do the following tasks:

- Calculate the correct amount of odorant to add to the gas stream
- Safely inject the odorant into the distribution line
- Monitor the odorant
- Calibrate the odorization equipment
- Generate reports regarding the odorization operation

11.0.0 ◆ PNEUMATIC SYMBOLS

Pneumatic symbols are used on blueprints of pneumatic systems. These symbols are a shorthand way of showing the location of items in the system, such as compressors and valves (see *Figure 33*). The pipeline mechanic, maintenance technician, and engineer use these symbols to plan, install, locate, troubleshoot, and maintain the system.

12.0.0 ◆ PIPELINE DESIGN

Pipeline used for transmission lines is made of carbon steel that meets standards set by the American Petroleum Institute. Distribution pipelines generally are made of plastic, but the material used may vary depending on location. Pipes fall into two broad size classifications: large-diameter piping (generally between 16 and 48 inches in diameter) and small-diameter piping (which can be as small as 0.05 inches in diameter). Large-diameter pipes, used in transmission lines, are formed by rolling the steel into a tube and welding the edges. Small-diameter pipes, used mainly in gathering and distribution lines, can be made seamlessly. To produce seamless piping, a hole is punched through a heated metal bar to create a tube.

Piping must be strong enough to endure many forces. On the inside, the pipeline must hold up under the heat and pressure from the flowing, compressed gas. Some lines also must be treated to rid them of bacteria that can form harmful acid deposits. Once a pipeline is buried, however, it is impossible for personnel to inspect the interior without digging it up again. Therefore, pipeline companies use sophisticated robots called pigs that can travel through the pipeline to inspect it. Pigs can perform the following tasks:

- Check the pipe wall thickness
- Detect corrosion

> **NOTE**
>
> Corrosion is the degradation of a material (in this case carbon or low alloy steel) through an interaction with its environment (called an electrolyte). Two electrochemical reactions are necessary for corrosion to occur: an oxidation reaction commonly called an anodic reaction and a reduction reaction commonly called a cathodic reaction.
>
> For corrosion on a pipeline to occur, the following four conditions must be present:
>
> - There must be an anode
> - There must be a cathode
> - There must be a metallic path to connect the anode and cathode (the pipeline itself can serve as this metallic path)
> - There must be an electrolyte, such as moist soil

- Detect leaks
- Look for any problem that could interrupt gas flow
- Look for any problem that could cause a safety risk

On the outside, the pipeline must withstand the effects of weather, erosion, and corrosion. A special epoxy coating that fuses to the pipeline is applied to protect it from corrosion. In addition, **cathodic protection** is used on some lines.

The following material summarizes the government regulations that cover pipeline design. For specific design details, it is best to refer to *49 CFR 192, Subpart C*. The following pipeline design factors are covered by this regulation:

- **Specified minimum yield strength**
- **Nominal** outside pipe diameter (see *Appendix*)
- Nominal pipe wall thickness (see *Appendix*)
- Design factor, a number based on the pipeline's geographic location
- Longitudinal joint factor, a number based on the **American Society for Testing and Materials (ASTM)** classification and pipe class (see *Table 1*)
- Temperature **derating** factor, a number based on the gas temperature (see *Table 2*)

Each of these factors is considered in a formula used by pipeline design engineers to calculate design pressure in pounds per square inch gauge.

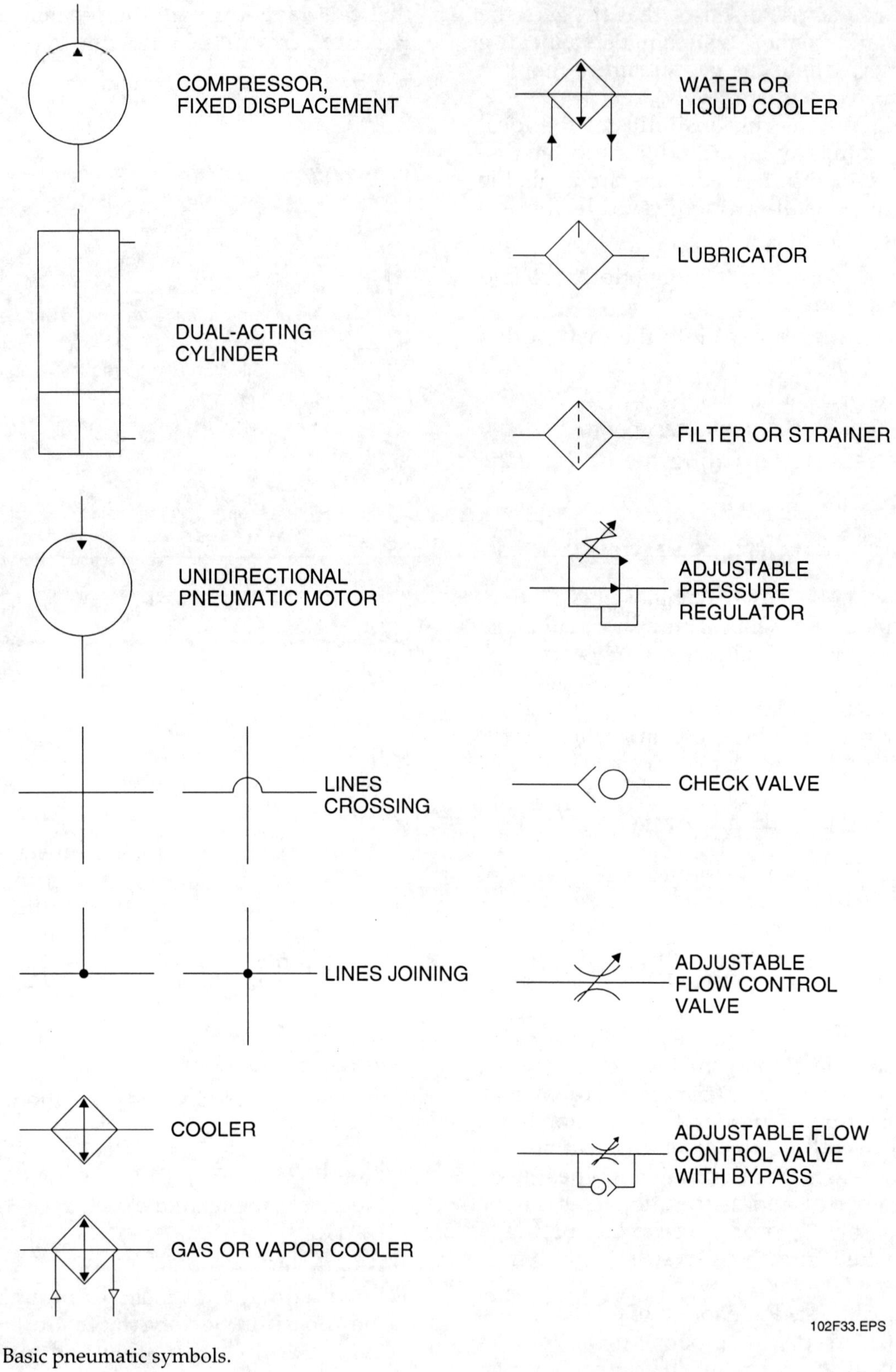

Figure 33 ◆ Basic pneumatic symbols.

Table 1 Longitudinal Joint Factor

ASTM Specification	Pipe Class*	Joint Factor
ASTM A 53	Seamless	1.00
ASTM A 53	Electric resistance welded	1.00
ASTM A 53	Furnace butt welded	0.60
ASTM A 106	Seamless	1.00
ASTM A 333/A 333M	Seamless	1.00
ASTM A 333/A 333M	Electric resistance welded	1.00
ASTM A 106	Seamless	1.00
ASTM A 381	Double welded	1.00
ASTM A 671	Electric fusion welded	1.00
ASTM A 672	Electric fusion welded	1.00
ASTM A 691	Electric fusion welded	1.00
API 5-L	Seamless	1.00
API 5-L	Electric resistance welded	1.00
API 5-L	Electric flash welded	1.00
API 5-L	Submerged arc welded	1.00
API 5-L	Furnace butt welded	0.60
Other	Pipe more than 4 inches (102 millimeters)	0.80
Other	Pipe 4 inches (102 millimeters) or less	0.60

* The type of longitudinal joint is determined by whether the pipe is welded and the type of weld used. If the type of longitudinal joint cannot be determined, the joint factor used must not be greater than that shown for "Other."
ASTM—The American Society for Testing and Materials
API—The American Petroleum Institute
Source: *49 CFR 192.113*

Table 2 Temperature Derating Factor

Gas Temperature in Degrees Fahrenheit (Celsius)*	Derating Factor
250°F (121°C) or less	1.000
300°F (149°C)	0.967
350°F (177°C)	0.933
400°F (204°C)	0.900
450°F (232°C)	0.867

* For intermediate gas temperatures, design engineers use interpolation to find the derating factor.
Source: *49 CFR 192.115*

12.1.0 Maximum Allowable Operating Pressure (MAOP)

The **maximum allowable operating pressure (MAOP)** for pipelines that carry natural gas is defined by government regulation *49 CFR 192.619.* If pipeline pressure goes higher than the lowest pressure that the weakest part of the pipeline can handle, the pipeline will fail. This regulation is designed to ensure that pipelines do not fail under operating pressure. It covers the following areas:

- Type of piping material used in the pipeline industry
- Areas where natural gas is transported by pipeline (onshore and offshore)
- Factors used by the pipeline industry to determine the MAOP

The MAOP of a pipeline is not a set number. It can change for any of the following reasons:

- Age of the pipeline
- Corrosion and wear
- Damage caused by nature
- Changes in federal and state regulations
- Changes in pipeline technology

> **NOTE**
> Pipelines can handle tremendous amounts of pressure. However, a pipeline is only as strong as its weakest part.

Changes in the material used to manufacture piping and system fittings (valves, flanges, and gaskets) may allow an increase in MAOP. For example, higher grades of steel, thicker pipe walls, and smaller diameter piping can all increase MAOP. However, there are some drawbacks. Thicker pipes are also heavier, which can increase transportation and construction costs. Friction may be greater in smaller diameter pipes, which can slow flow and affect efficiency.

The American National Standards Institute (ANSI) determines the grades of piping and fittings to be used in pipeline construction. ANSI is a private, nonprofit organization that

- Works with qualified industry volunteers to set standards for a variety of industries
- Maintains standards records
- Publishes current industry standards

12.2.0 Hydrostatic Testing

Before the pipeline is put into service, it must be tested to ensure that it will not leak. Pipeline companies are required by law to test for leaks for the following reasons:

- To protect lives and property
- To protect the environment
- To conserve energy

The safest and most economical way to do this is with a hydrostatic test. To start the test, water is pumped into the line and put under pressure for a certain amount of time. The amount of pressure and the length of the test depend on several factors, including piping material, size, and the eventual operating pressure of the line.

A significant drop in pressure during the test indicates a leak somewhere in the line. To find the exact location of the leak, testing personnel rely on special paint that has been applied to welds and joints. This paint is designed to permanently change color when it comes into contact with water. Thus, even if the drop in pressure stops a leak before the test is finished, workers can still see where the problem is.

Government regulations also require pipeline companies to keep a record of any tests performed on the line (*49 CFR 192.517*). The record must contain the following information:

- Pipeline company name
- Names of those responsible for conducting the test
- Test medium used
- Test pressure
- Length of the test
- Pressure recording charts
- Elevation variations if appropriate
- Leaks and failures noted and how they were dealt with

In addition, government regulations require pipeline companies to conduct the tests safely. Every reasonable effort must be made to protect employees, the public, and the environment from the potential hazards associated with testing.

Review Questions

Sections 11.0.0–12.0.0

1. _____ are a shorthand way of showing the location of items in the system, such as compressors and valves.
   a. Digital prints
   b. Pneumatic symbols
   c. Component layouts
   d. Schematic grids

2. Large-diameter piping is generally _____ in diameter.
   a. between 16 and 48 inches in diameter
   b. between 48 and 52 inches in diameter
   c. no less than 60 inches in diameter
   d. no less than 80 inches in diameter

3. All of the following must be present for pipeline corrosion to take place except _____.
   a. an anode
   b. a cathode
   c. a metallic path connecting the anode and cathode
   d. dry soil

4. The _____ determines the grades of piping and fittings to be used in pipeline construction.
   a. Code of Federal Regulations (CFR)
   b. American Petroleum Institute (API)
   c. American National Standards Institute (ANSI)
   d. American Society for Testing and Materials (ASTM)

5. The safest and most economical way to test a pipeline for leaks before putting it into service is a ____ test.
   a. strength
   b. hydrostatic
   c. nominal wall thickness
   d. corrosion

Summary

When natural gas is released from a well, it is under a lot of pressure. That pressure produces forward movement. However, as gas enters the pipeline, whether from a well or the refinery, friction causes it to lose pressure and slow down. To keep the gas moving at acceptable operating levels, compressor stations located at intervals along the pipeline compress the gas again.

In this module, you learned about the study of the mechanical properties of gases, which is called pneumatics. You also were introduced to the pneumatic and other equipment used on the pipeline. Finally, you were given a basic background in the principles of pipeline design.

The pipeline might be considered the most important part of the natural gas system. It is the connection between the source of natural gas and the customers who depend on it every day. Therefore, it must be built to, tested to, and maintained at the highest standards.

GLOSSARY

Trade Terms Introduced in This Module

Absorption: The penetration of a substance into the body of another.

Actuator: A mechanism that indirectly moves or controls a part.

Adsorption: The condensation of gases, liquids, or dissolved substances on the surfaces of solids.

American Society for Testing and Materials (ASTM): A nonprofit organization that works with volunteers in the industry to set standards.

Cathodic protection: A method used to prevent the destructive electrochemical process of corrosion of metal by using a sacrificial anode.

Compression ratio: The ratio of the absolute discharge pressure to the absolute intake pressure. The ratio usually applies to a single stage of compression, but it may be applied to multiple stages of compression.

Compressor: A machine that compresses natural gas to reduce its volume and force it through the pipeline.

Control center: The operations center of a pipeline, usually located in a terminal or main pipeline facility. All the control consoles and computer-operated systems are located in the control center.

Custody transfer: The point at which a change of ownership or responsibility for the natural gas in a pipeline takes place.

Derating: The reduction of the rating of a device to improve reliability or to permit operation at high ambient temperatures.

Diffuser: A duct, chamber, or section in which a high-velocity, low-pressure stream of air or other gas is converted into a low-velocity, high-pressure flow.

Discharge: A term that refers to the area where the gas leaves a compressor. It is sometimes called the outlet.

Discharge pressure: The total gas pressure at the discharge flange of the compressor.

Discharge temperature: The total gas temperature at the discharge flange of the compressor.

Distribution line: A pipeline that transports gas from a common source of supply to (1) a customer's meter or the connection to a customer's piping, whichever is furthest downstream or (2) the connection to a customer's piping if there is no customer meter (*49 CFR 192.3*).

Downstream: Any point farther beyond the origination than a given point.

Gathering line: A pipeline that transports gas from a current production facility to a transmission line (*49 CFR 192.3*).

Lockout/tagout: A formal procedure that consists of two steps: locking a piece of equipment that must be serviced and placing a tag on that equipment to alert employees that the equipment must not be used. Only qualified employees may lock out and tag out equipment, certify that equipment is ready for use, and remove the locking device and tag.

Material safety data sheet (MSDS): A form containing information on a potentially hazardous material, including variations on the material's name, contents, test results, characteristics, potential hazards, and handling procedures.

Maximum allowable operating pressure (MAOP): The highest level of internal gauge pressure allowed for a particular pipeline or pipeline segment, as set by the U.S. Department of Transportation.

Nominal: A closely approximate measurement rather than an actual measurement. For example, a 2 × 4 (the nominal size) piece of wood actually measures 1⅝ × 3⅝ inches.

Odorant: A liquid added to natural gas (which is odorless) as a safety feature to make it smell like rotten eggs.

Poppet: A valve that rises straight up to or from its seat.

Seat: The nonmoving part of a valve upon which the moving part rests to form a seal and close the valve.

Shut in: To isolate a segment of the gas pipeline or a pipeline facility associated with that segment in response to an abnormal operating condition (AOC) or to make repairs.

Solenoid: A coil of wire shaped into a long cylinder. A moveable core is drawn into the coil when current flows.

Specified minimum yield strength (SMYS): The specified minimum yield strength of steel pipe, expressed in pounds per square inch (psi) as defined in Department of Transportation regulations.

Stuffing box: A device that prevents leaks along a moving part, such as a piston rod.

Suction: A term that refers to the area where the gas enters a compressor. It is sometimes called the inlet.

Suction pressure: The total pressure of a gas or fluid entering the suction valve of a compressor.

Torque: A force that produces rotation, turning, or twisting.

Transmission line: A pipeline that transports gas from a gathering line or storage facility to a distribution center, storage facility, or large-volume customer that is not downstream from a distribution center (*49 CFR 192.3*).

Upstream: Any point located closer to the origination than a given point.

Valve: A device used to control the rate of flow in a pipeline, open or shut off a pipeline, or serve as a safety device.

APPENDIX

Pipe Wall Thickness

Dimensions of Welded and Seamless Pipe
Carbon, Alloy and Stainless Steel
ASME B36.10, B36.19

Nominal Pipe Size (In Inches)	Outside Diameter	Sch 5S	Sch 10S	Sch 10	Sch 20	Sch 30	Sch 40S	STD	Sch 40	Sch 60	Sch 80S	XS	Sch 80	Sch 100	Sch 120	Sch 140	Sch 160	XXS
⅛	10.29	–	1.24	–	–	–	1.73	1.73	1.73	–	2.41	2.41	2.41	–	–	–	–	–
¼	13.72	–	1.65	–	–	–	1.73	1.73	1.73	–	3.02	3.02	3.02	–	–	–	–	–
⅜	17.14	–	1.65	–	–	–	2.31	2.31	2.31	–	3.20	3.20	3.20	–	–	–	–	–
½	21.34	1.65	2.11	–	–	–	2.77	2.77	2.77	–	3.73	3.73	3.73	–	–	–	4.75	7.47
¾	26.67	1.65	2.11	2.11	–	–	2.87	2.87	2.87	–	3.91	3.91	3.91	–	–	–	5.54	7.82
1	33.40	1.65	2.77	2.77	–	–	3.38	3.38	3.38	–	4.55	4.55	4.55	–	–	–	6.35	9.09
1¼	42.16	1.65	2.77	2.77	–	–	3.56	3.56	3.56	–	4.85	4.85	4.85	–	–	–	6.35	9.70
1½	48.26	1.65	2.77	2.77	–	–	3.68	3.68	3.68	–	5.08	5.08	5.08	–	–	–	7.14	10.16
2	60.32	1.65	2.77	2.77	–	–	3.91	3.91	3.91	–	5.54	5.54	5.54	–	–	–	8.71	11.07
2½	73.02	2.11	3.05	3.05	–	–	5.16	5.16	5.16	–	7.01	7.01	7.01	–	–	–	9.52	14.02
3	88.90	2.11	3.05	3.05	–	–	5.49	5.49	5.49	–	7.62	7.62	7.62	–	–	–	11.13	15.24
3½	101.60	2.11	3.05	3.05	–	–	5.74	5.74	5.74	–	8.08	8.08	8.08	–	–	–	–	16.15
4	114.30	2.11	3.05	3.05	–	–	6.02	6.02	6.02	–	8.56	8.56	8.56	–	11.13	–	13.49	17.12
5	141.30	2.77	3.40	3.40	–	–	6.55	6.55	6.55	–	9.53	9.53	9.53	–	12.70	–	15.88	19.05
6	168.28	2.77	3.40	–	–	–	7.11	7.11	7.11	–	10.97	10.97	10.97	–	14.27	–	18.24	21.95
8	219.08	2.77	3.76	–	6.35	7.04	8.18	8.18	8.18	10.31	12.70	12.70	12.70	15.06	18.26	20.62	23.01	22.22
10	273.05	3.40	4.19	–	6.35	7.80	9.27	9.27	9.27	12.70	12.70	12.70	15.06	18.26	21.44	25.40	28.58	25.40
12	323.85	3.96	4.57	–	6.35	8.38	9.52	9.52	10.31	14.27	12.70	12.70	17.48	21.44	25.40	28.58	33.32	25.40
14	355.60	3.96	4.78	6.35	7.92	9.52	–	9.52	11.13	15.06		12.70	19.05	23.83	27.79	31.75	35.71	–
16	406.40	4.19	4.78	6.35	7.92	9.52	–	9.52	12.70	16.66		12.70	21.44	26.19	30.96	36.52	40.46	–
18	457.20	4.19	4.78	6.35	7.92	11.12	–	9.52	14.27	19.05		12.70	23.82	29.36	34.92	39.67	45.24	–
20	508.00	4.78	5.54	6.35	9.52	12.70	–	9.52	15.06	20.62		12.70	26.19	32.54	38.10	44.45	49.99	–
22	558.8	4.78	5.54	6.35	9.52	12.70	–	9.52	15.87	22.22		12.70	28.58	34.92	41.28	47.62	53.98	–
24	609.4	5.54	6.35	6.35	9.52	14.27	–	9.52	17.48	24.61		12.70	30.93	38.89	46.02	52.37	59.51	–
26	660.4	–	–			–	–	9.52	–	–		12.70	–	–	–	–	–	–
28	711.2	–	–	7.92	12.70	15.88	–	9.52	–	–		12.70	–	–	–	–	–	–
30	762.0	6.35	7.92	7.92	12.70	15.88	–	9.52	–	–		12.70	–	–	–	–	–	–
32	812.8	–	–	7.92	12.70	15.88	–	9.52	17.48	–		12.70	–	–	–	–	–	–
34	863.6	–	–	7.92	12.70	15.88	–	9.52	17.48	–		12.70	–	–	–	–	–	–
36	914.4	–	–	7.92	12.70	15.88	–	9.52	19.05	–		12.70	–	–	–	–	–	–
38	965.2	–	–	–	–	–	–	9.52	–	–		12.70	–	–	–	–	–	–
40	1,016.0	–	–	–	–	–	–	9.52	–	–		12.70	–	–	–	–	–	–
42	1,066.8	–	–	–	–	–	–	9.52	–	–		12.70	–	–	–	–	–	–
44	1,117.6	–	–	–	–	–	–	9.52	–	–		12.70	–	–	–	–	–	–
46	1,168.4	–	–	–	–	–	–	9.52	–	–		12.70	–	–	–	–	–	–
48	1,219.2	–	–	–	–	–	–	9.52	–	–		12.70	–	–	–	–	–	–

102A01.EPS

Figure Credits

Instromet, Inc.	102F28
Emerson Process Management, Daniel Division	102F29, 102F30, 102F31
Fisher Controls International, LLC	102F32
The American Society of Mechanical Engineers	Appendix A

CONTREN™ LEARNING SERIES — USER UPDATES

The NCCER makes every effort to keep these textbooks up-to-date and free of technical errors. We appreciate your help in this process. If you have an idea for improving this textbook, or if you find an error, a typographical mistake, or an inaccuracy in NCCER's Contren™ textbooks, please write us, using this form or a photocopy. Be sure to include the exact module number, page number, a detailed description, and the correction, if applicable. Your input will be brought to the attention of the Technical Review Committee. Thank you for your assistance.

Instructors – If you found that additional materials were necessary in order to teach this module effectively, please let us know so that we may include them in the Equipment/Materials list in the Instructor's Guide.

Write: Curriculum Revision and Development Department
National Center for Construction Education and Research
P.O. Box 141104, Gainesville, FL 32614-1104

Fax: 352-334-0932

E-mail: curriculum@nccer.org

Craft ____________________ Module Name ____________________

Copyright Date __________ Module Number __________ Page Number(s) __________

Description

____

____

____

____

(Optional) Correction

____

____

____

(Optional) Your Name and Address

____

____

____

Module 67103-02

Pipeline Communications

Introduction to the Pipeline Operator Qualification Rule, API, and NCCER

It could be said that petroleum products keep America running. Petroleum products like oil and natural gas are the principal sources of energy in the United States and are used in many manufacturing and chemical processes. Petroleum products are primarily transported by pipeline. There are more than a million miles of liquid and natural gas pipeline in this country, ranging in diameter from two inches to four feet. These pipelines bring crude oil and natural gas from production fields to processing plants and refineries—and ultimately to the consumer.

Pipeline Operator Qualification

Petroleum products are considered hazardous, so pipeline personnel must take extra precautions and follow specific procedures when working with them. To help ensure the safe handling and transportation of petroleum products, the United States Department of Transportation (DOT) mandates safe practices as federal regulations.

One such regulation, the Operator Qualification (OQ) rule, requires pipeline operators to develop a written qualification plan to evaluate their personnel and their contract personnel (see *49 CFR Part 192 Subpart N—Gas*, and *49 CFR Part 195 Subpart G—Liquid*). The evaluation hinges on the operator's ability to perform certain covered tasks and to be able to recognize and respond to abnormal operating conditions that may be encountered while performing these covered tasks.

Covered Tasks

A pipeline performance task is deemed *covered* if the activity passes the following four-part test:

1. Is it performed on a pipeline facility?
2. Is it an operations or maintenance task?
3. Is it performed as a requirement of *49 CFR Part 195 (Liquid)* or *49 CFR Part 192 (Gas)*?
4. Does it affect the operation or integrity of the pipeline facility?

If the answer to all four questions is *yes*, then the activity is considered *covered* and must figure into the evaluation process of pipeline personnel.

American Petroleum Institute

The OQ rule and its related covered tasks provided the catalyst to develop a unified pipeline training and assessment program—so that pipeline operators and owners have ready access to the assessment and training tools they need. This effort was the direct result of the vision and leadership provided by The American Petroleum Institute (API) and API members.

Efforts to unify the industry are nothing new to API. Indeed, the impetus for forming the Institute in 1919 was the need to standardize engineering specifications for drilling and production equipment. Today, the most pressing issues revolve around public perceptions and government policies toward the industry. Speaking with one voice on these issues has become as essential as having interchangeable parts in the field. API's members determine the petroleum industry's positions on public policy as well as the standards that govern its day-to-day operations worldwide.

Consortium on Operator Qualification

API assembled a focus group whose primary purpose was to identify covered tasks and provide compliance recommendations to pipeline operators.

The Consortium on Operator Qualification (COOQ) became the guiding force for the identification and development of covered tasks for the liquid pipeline industry. Their work product, API publication 1161, *Guidance Document for the Qualification of Liquid Pipeline Personnel*, formed the basis for the resulting Pipeliner Training and Assessment Program.

The covered task list created in API publication 1161 has now evolved to include gas covered tasks and is maintained by NCCER. The most current covered task list can be downloaded at www.nccer.org.

National Center for Construction Education and Research

With the guidance and support of API, its members, and the Consortium on Operator Qualification, the National Center for Construction Education and Research (NCCER) was able to design and implement the Pipeliner Training and Assessment Program.

NCCER is a not-for-profit 501(c)(3) education foundation established in 1995 by the world's largest and most progressive construction and maintenance contractors and national contractor associations. NCCER was founded to address the

severe workforce shortage facing the construction and maintenance industries and to develop a standardized training process and curricula. Today, NCCER is supported by hundreds of leading construction and maintenance companies, manufacturers, and national associations and developed more than 35 construction and maintenance titles.

NCCER also maintains a National Registry, which contains the records of trainees and instructors who have received NCCER training. All completion information for every trainee is sent to the NCCER and filed within the Registry. The Registry can then confirm training and skills for workers as they move from company to company, state to state, or even within their own company.

Standardized Curriculum (NCCER's Contren™ Learning Series)

NCCER curricula are developed using the expertise of industry subject matter experts (SMEs). The use of SMEs as an integral part of every new development or revision ensures that the curriculum is current, industry-driven, and indicative of industry practices. The development and/or revision process is facilitated from start to finish by NCCER project managers, who work closely with the SMEs and a select team of technical writer(s) on the development and presentation of the curriculum.

Pipeline: Five Training Paths

The pipeline curriculum focuses on essential skills and competencies within five skill areas:

- Corrosion Control
- Electrical and Instrumentation
- Maintenance
- Mechanical
- Operations (Field and Control Center)

SMEs from the pipeline industry have determined the skills and competencies needed based on API's covered tasks and the consensus of the group/industry. Specific company information is generally not incorporated into the curriculum unless the SMEs determine that the information is the industry standard.

Training Modules and Covered Tasks

The smallest unit of instruction in the curriculum is a *module*. Each curriculum is composed of modules (the equivalent of stand-alone chapters) that are logically sequenced into complete *levels* of instruction. Every module is task-oriented and requires the successful completion of a written (knowledge-based) examination and a performance (hands-on) test. Modules may be individually selected and/or combined for efficient task training. Modules in the pipeline curricula, to the extent possible, will align with API-developed covered tasks. While not all modules will relate to covered tasks, all covered tasks will relate to at least one module.

Benefits of Training to the Pipeline Marketplace

Applying NCCER's standardized training process allows pipeline owners and contractors in the United States to establish the benchmark for acceptable skill and competency levels within the pipeline industry. This process will assist training organizations in developing and/or improving standardized pipeline training programs that effectively match their resources and operating conditions while responding to regulatory demands and to the needs of the industry. The process will also recognize training organizations that meet the minimum standards of competence, warranting regulatory, public, and professional confidence. This in turn will foster national unity within the pipeline industry for the mutual benefit of individual technicians, owners, contractors, and ultimately pipeline-users and society.

Additional Resources

To review the Pipeliner Training and Assessment Program Implementation and Resource Guide and the latest version of pipeline covered tasks, visit NCCER's Web site at www.nccer.org, or contact:

National Center for Construction Education and Research
PO Box 141104
Gainesville, FL 32614-1104
(888) 622-3720
Web site: www.nccer.org

For more information on API, please contact:

American Petroleum Institute
1220 L Street, Northwest
Washington, D.C. 20005
(202) 682-8000
Web site: www.api.org

To order *API Pub 1161*, contact Global Engineering Documents by calling 1-800-854-7149 or visiting their Web site at www.global.his.com.

Additional sources of information on the pipeline operator qualification regulation include:

DOT – Office of Pipeline Safety
www.ops.dot.gov

Transportation Safety Institute
www.tsi.dot.gov

COURSE MAP

This course map shows all of the modules in the Gas Pipeline Operations curriculum. The suggested training order begins at the bottom and proceeds up. Skill levels increase as you advance on the course map. The local Training Program Sponsor may adjust the training order.

GAS PIPELINE OPERATIONS

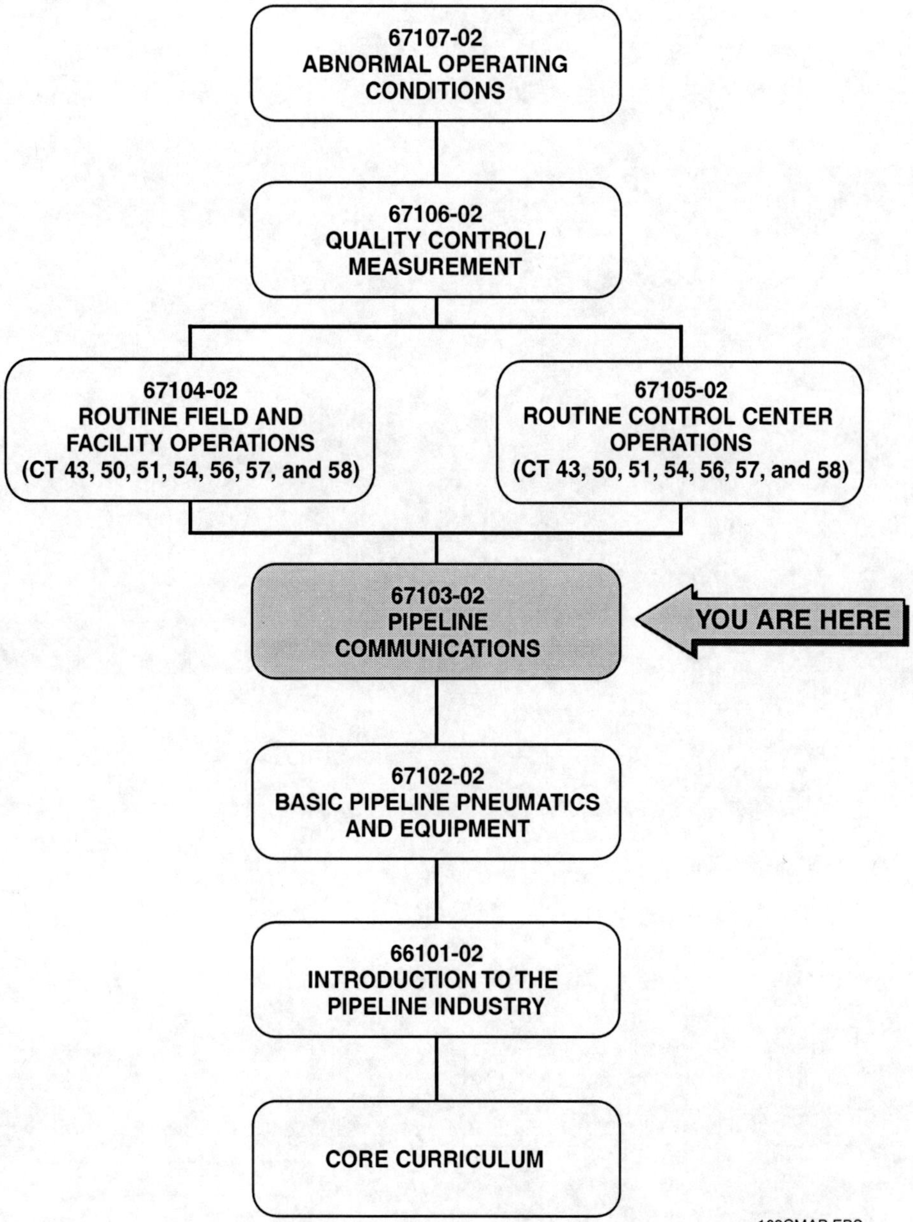

103CMAP.EPS

MODULE 67103-02 CONTENTS

Figures

MODULE 67103-02

Pipeline Communications

Objectives

When you have completed this module, you will be able to do the following:

1. Communicate critical operations information to the next shift.
2. Communicate with the control center, scheduling department, and maintenance technicians.
3. Initiate call-outs.
4. Generate work orders.
5. Effectively communicate with pipeline personnel.
6. Notify customers of operations changes.
7. Effectively communicate with the public regarding pipeline issues.

Prerequisites

Before you begin this module, it is recommended that you successfully complete the following: Core Curriculum; Pipeline Core, Module 66101-02; Gas Pipeline Operations, Module 67102-02.

1.0.0 ◆ INTRODUCTION

A compressor stops working during one shift, but that information is not given to the next shift. A maintenance crew arrives to do repairs, but no one has told the **field** workers about this interruption to their work. A field supervisor must be called back in to solve a problem at 1:00 a.m., but the contact numbers on file are not correct, and the supervisor cannot be reached. These are all examples of poor communication. The result of poor communication is usually frustration and lost time. However, poor communication can have more serious results. A breakdown in communication on the pipeline could also result in any of the following problems:

- A dangerous condition resulting in injuries to or death of pipeline employees or the public
- Serious damage to expensive pipeline equipment
- Harm to the environment
- Loss of income to the pipeline company
- Poor service to customers who depend on safe, reliable delivery for heating, cooking, and industrial processes

Accurate communication is vital to pipeline **operations.** It is so important that it is covered in federal laws and in company policy manuals. In this module, you will get a broad overview of how communication works in the pipeline industry. Note that each company has its own procedures and forms. Therefore, you must learn and follow your company's policy.

2.0.0 ◆ INTERNAL COMMUNICATIONS

Internal communications should work like the pipeline at its best: smoothly, clearly, and efficiently. The work you do affects the work performed by your co-workers and vice versa, so it is important to share information correctly and in a timely fashion. Examples of this type of information sharing follow:

- Giving operations information to the next shift
- Giving information to the control center
- Coordinating maintenance work
- Issuing **call-outs**
- Issuing **work orders**

Many of the processes on the pipeline are controlled by the **Supervisory Control and Data Acquisition (SCADA)** computer system. Information about the pipeline system and its equipment is collected, monitored, analyzed, and stored in this system in the control center. However, spoken and written communication among employees is also important.

2.1.0 Communication between Shifts

Many pipeliners work in shifts. Therefore, if a problem comes up on one shift, the next shift must be told about it. Good communication between shifts is important for the following reasons:

- To protect worker safety
- To protect pipeline equipment
- To avoid costly downtime
- To prevent actions being taken based on poor or missing information
- To provide good customer service

Shift workers write down instructions for the next shift in a logbook. Sometimes instructions are given orally. In either case, workers must ensure that their instructions are clear and fully understood by the next shift. To ensure that instructions are understood, workers must describe exactly what the problem is and where it is located. They may also include the following information:

- The fact that a repair crew is scheduled during the shift
- Information from scheduling that could affect operations
- Weather conditions that could affect the field workers

Employees may be seriously injured or killed by hazardous energy released during repair and maintenance of machines and equipment. To ensure their safety, pipeline companies use lockout/tagout procedures. Lockout is the practice of locking equipment that requires repair or routine maintenance. The locks prevent employees from accidentally operating such equipment until it is safe to do so. Tags placed on locked equipment alert employees that the equipment must not be used until the locks and tags are removed. Locks and tags may be placed or removed only by trained, authorized employees. These employees must follow the lockout/tagout standards established by the Occupational Safety and Health Administration (*29 CFR 1910.47*) and company policies and procedures.

2.2.0 Control Center Communications

Pipeline operations are coordinated between the field and the control center. Each group stays in touch so that everyone knows what is happening on the pipeline. Generally, almost all information is collected and communicated through the SCADA computer system in the control center (see *Figure 1*). However, field workers also send information that may not be available from that system to the control center. Communications between the control center and the field are sent to keep the pipeline within proper operating limits and to monitor the pipeline for problems.

103F01.EPS

Figure 1 ◆ Pipeline control center.

2.3.0 Coordinating Maintenance Work

Pipeline maintenance is necessary. However, the pipeline operates around the clock every day of the year, and maintenance work may interrupt some operations. The field maintenance supervisor must ensure that interruptions are as short as possible. This supervisor must also ensure that the work is performed as safely as possible. The supervisor meets these goals by doing the following tasks:

- Notifies all affected workers to let them know what work must be performed, where it will be performed, and how long it will take
- Makes regular progress reports while the work is ongoing and reports any expected delays
- Checks on the work to keep it on schedule
- Checks to ensure that all work is complete
- Tells the field and control center when the pipeline is back in service

The control center keeps tabs on the work to ensure that no other problems come up while the work is being performed. During maintenance work, the control center employees perform these tasks:

- Make operating changes and record changes on the **daily operating sheet**
- Monitor pipeline pressure
- Keep the field informed about changes in pipeline pressure

Maintenance workers must work quickly to return the line to normal operations as soon as possible. However, they must also work safely and carefully. They must try to avoid these common errors:

- Opening the wrong valve
- Bypassing a safety device
- Operating equipment incorrectly

2.4.0 Issuing Call-Outs

Sometimes pipeliners are called to work outside their normal working hours. This is referred to as a call-out. Workers are called out to deal with problems that must be handled immediately. Most pipeliners carry cell phones or pagers so that they can be located quickly. You must give your company accurate telephone, cell phone, or pager numbers and keep this information up to date so that you can be found when needed.

There are two basic types of call-out events. The first type involves a problem that must be fixed immediately but does not require outside help. Pipeliners can handle this type of problem safely using internal resources. The second type of call-out is a significant emergency that does require outside help. Note that emergency response agencies are not the only groups that may become involved during a major emergency. Such emergencies may involve representatives from all or some of the following groups:

- Local emergency response agencies (fire department, law enforcement, U.S. Coast Guard)
- Other pipeline **operators** (by agreement)
- Public facilities (transportation, communication)
- Local government
- The media

You must follow company policy for contacting and working with outside organizations. More information about communicating with outside organizations is presented elsewhere in this module.

2.5.0 Issuing Work Orders

Work orders are written instructions that describe the work to be completed and who should do it. Only authorized employees may issue work orders. Work orders are issued for general maintenance and repair work and for regularly scheduled preventive maintenance. In most cases, preventive maintenance must be performed on a schedule required by government regulations. Pipeliners use computer programs to track maintenance schedules, issue work orders, and record completed work. Work orders may include the following information:

- Work order number
- Control center location
- Description of the work needed
- Location where the work will be performed
- Date and time for the work to begin
- Date and time for the work to be completed
- Name of the supervisor or lead technician
- Signature of the control room supervisor and maintenance supervisor

> **NOTE**
> Some work orders are issued long before the date when the work will actually start. Such long-range work orders usually involve a large project that will have a major impact on operations. Therefore pipeline operators must plan carefully to ensure that the project is completed safely, cost effectively, and on time.

Once the work is finished, more information must be recorded. The following information may be required after the work is completed:

- Control center location
- Shift times and dates for starting and completing the work
- Name of the supervisor in charge
- Name of the lead technician
- Name of the control center operator
- Work item number
- Description of the type of work and where it was performed
- List of the procedures followed
- Work permit or work order number
- Names of the people assigned to do the work
- Initials of the approving supervisor
- Day and time the work was completed
- Initials of the person checking the completed work
- Comments

The paperwork also includes permits that may be required by company policy or law. For example, a permit may be needed to allow entry into a confined space or to do electrical work. Copies of all required permits must be filed with the work orders.

2.6.0 Communications with Scheduling and Customers

Changes in pipeline operations can affect delivery of the gas. Appropriate personnel working in scheduling must be notified of problems that can affect their work. In addition, customers may need to be notified of any problems that could affect delivery.

2.6.1 Communications with Scheduling

When there are repairs or problems on the line, the schedule may need to be adjusted to keep the line in balance and to keep the gas flowing to customers. Therefore, the scheduling department must be kept informed about the condition and availability of pipeline equipment. Common reasons for adjusting the schedule include these items:

- Maintenance work
- Emergency repairs
- Requests for an increase or decrease in delivery amounts
- Requests for schedule verification
- Requests for product quality verification

2.6.2 Communications with Customers

Communications with customers involve scheduling deliveries, answering questions, and solving problems. Generally, the pipeline operator's customers include local distribution companies who purchase gas to supply it to **end users** and **marketers** who purchase gas supplies to resell them to local utilities or end users. The pipeline transmission company generally does not deal directly with end users. The local utility or local distribution company typically handles those communications.

Computers are used to manage all of the steps involved in the purchase and transmission of gas to customers. Costs are calculated based on the cost of the gas itself and the cost to transport the gas through the pipeline. Customers use the **nomination process** to purchase their gas supply and reserve time on the pipeline for transmission.

2.6.3 Nomination Process

Gas is moved through the pipeline in much the same way that trains are moved along tracks. The gas may flow directly from point A to point B within one state, or it may cross several state lines. The gas also may be diverted into another, less direct line or even be held for a time in a storage line.

A nomination is a request from a customer to move gas through the pipeline to a delivery point. Pipeline operators get nominations daily. Customers may request different volumes of gas at different times of the day or even change an order after transmission has started. It is a fairly complicated process, and pipeline operators use computers to handle it.

Before customers can take part in the nomination process, they must sign an agreement or contract with the pipeline company. Pipeline operators provide software so that customers can make nominations for their gas supply online. Sometimes customers can use fax or e-mail; however, most operators require more time to process such requests.

A typical nomination includes these items:

- Source of the gas
- Destination point
- Receipt and delivery points
- Proposed volume of the gas to be delivered
- Contract number
- Start and end date for the delivery

The pipeline operator uses the computer system to manage the entire nomination process. The system allows the operator to do the following tasks:

- Track orders
- Send gas directly from its source to its destination point
- Temporarily store gas (**line pack**) to make room on the main line for a higher priority order
- Change the order of deliveries to meet higher priority requests
- Adjust gas volume during transmission if a customer needs more or less gas than requested
- Manage gas inventory

The online nomination system also allows customers to manage their gas delivery needs. Customers use the software provided by the pipeline company to accomplish these tasks:

- Choose the source of gas
- Change the volume of gas ordered
- Review their account status

Pipeline operators must keep their customers informed about conditions that could affect delivery times and volumes. Such conditions may include imbalance between customer use and pipeline inventory and shortages. Notifications about conditions that could affect delivery are also available to customers online.

2.7.0 Communications with Quality Assurance

The quality assurance department tests and confirms product quality. Natural gas may contain other gases, such as carbon dioxide, nitrogen, hydrogen sulfide, and oxygen. The levels of these other gases must not exceed quality and safety standards.

Quality assurance personnel must inform the control center of any condition that could affect product quality. The control center tracks this information and informs the appropriate departments about any actions that may have to be taken. The results of quality tests, as well as any problems and how they were solved, are filed in company records.

2.8.0 Communications with Line Patrols

Usually patrols fly over the pipeline on a regular basis to check its condition. However, if bad weather or other conditions ground flights, the line can be checked by vehicle or on foot. When a problem is spotted, patrols radio or telephone their reports to personnel in the control center. Control center personnel must note the report number and the date, time, and location of the problem. Patrols may provide the location of problems by naming the closest milepost, providing the map coordinates, or using the global positioning system.

The operations department must alert a supervisor to any conditions that indicate a problem with the pipeline. The supervisor then sends a team to investigate. The team reports the results of its investigation to the control center, and a decision is then made about how to solve the problem.

Communication between line patrols and the control center must be clear and precise. Sounds can be distorted during radio, telephone, or cell phone reports. Many spoken words can sound alike. Therefore, to ensure that information is understood, both control center staff and patrols should use these tips to improve communication:

- If information is not understood, ask that it be repeated
- Spell out place names
- Restate information to confirm what was said
- Use exact descriptions

Using exact descriptions is the best way to communicate information about the condition of the line. For example, a patrol might state that the pipeline is in poor or bad condition, but what does that mean? Is the pipeline burnt, gouged, or ruptured? Is there erosion near the line? Employees who patrol the line must use the right words so that the control center understands exactly what is wrong.

3.0.0 ◆ EXTERNAL COMMUNICATIONS

The federal government requires pipeline operators to keep the public informed about their activities. The term *public* includes the general public, customers, businesses, emergency response agencies, the media, and government organizations. Following are some of the channels through which operators meet this responsibility:

- Annual reports
- Printed informational materials
- Web sites
- Press releases
- Advertisements
- Meetings and other personal contacts

The most common manner of communicating with the public is by telephone. When answering telephone calls, it is important to follow company policy. A call may be a routine matter, or it could be an alert about a dangerous situation. All callers must be treated with courtesy and respect.

3.1.0 Communicating with the General Public

Members of the general public may call the pipeline company to ask questions, to complain about a problem, or to tell the company about a dangerous situation. To ensure that calls are handled professionally, employees responsible for taking calls from the general public must follow these guidelines:

- Understand and follow company procedures for handling inquiries and reports
- Use the appropriate company form to make a record of the call
- Assure callers that their inquiry will be referred to the appropriate company representative
- Refer the call to the appropriate company representative

Depending on company policies and procedures, the form used to make a record of calls may be a standard message form or an investigation report. An example of a Call Investigation Report is shown in *Figure 2*.

3.2.0 Educating the Public

To promote safety, the federal government requires pipeline operators to set up public education programs (*49 CFR 192.616*). Such programs help the public to do the following:

- Recognize pipeline **right-of-way (ROW)** markers
- Recognize a gas pipeline emergency
- Know what to do in case of a gas leak
- Learn about pipeline operations
- Learn about the **One-Call notification system** and how to use it
- Learn how to protect themselves in an emergency

Pipeline operators must ensure that everyone in their service area has access to educational materials. For example, most printed materials are in English. However, the company's service area may include a large number of non-English-speaking people. In such cases, federal laws state that the company must provide educational materials in other languages as needed.

As part of their public education program, pipeline operators perform some basic tasks, including these examples:

- Producing and giving out educational material
- Holding meetings
- Putting up ROW markers
- Reviewing and auditing the educational program

3.2.1 Informational Material

Pipeline operators provide a wide variety of printed informational materials to the public, including those listed here:

- A description of the pipeline system, including a system map
- Color photographs showing what pipeline markers look like
- A description of pipeline marker signs that tells what information is printed on the markers and where they are usually located
- An explanation of how to identify the location of pipelines
- Reminders to call the One-Call notification system before digging, excavating, or plowing near the pipeline, which may be printed on bills or small items such as pencils or notepads given to the public
- A list of One-Call notification system telephone numbers
- The company's toll-free emergency telephone number
- An explanation of how to recognize a pipeline release
- A list of actions to take in case of a pipeline release
- An explanation of the company's damage prevention program

Generally, the local distribution company is responsible for providing information about the odorant that is added to natural gas as a safety feature. The company may provide this information in utility bills, through recorded information on its telephone answering system, on its Web site, or in informational mailings to its customers. Printed information may include the following information:

- How to recognize the smell of gas (some companies use a scratch-and-sniff card)
- What to do if the smell of gas is detected
- What not to do if the smell of gas is detected

3.2.2 Meetings

Pipeline operators sometime use meetings to educate the public. Usually, these meetings are held for specific groups. Meetings are an effective way to personally educate the public, to answer specific questions, and to hear concerns. Federal laws state that the company must provide a translator if needed so that non-English-speaking people may also hear the information. In addition, companies may be required to provide sign-language interpreters if necessary.

3.2.3 Right-of-Way Markers

Right-of-way markers are signs that alert people that a pipeline is buried underground. To prevent damage to the pipeline, a line marker must be placed over each buried main and transmission line at each crossing of a public road and railroad and wherever needed to identify the location of the line.

Line markers generally are not required at main and transmission lines located either offshore or at crossings of or under bodies of water.

Federal laws also require that enough markers be located along buried pipeline to clearly show where the pipeline is. As is often the case with federal laws, there are some exceptions. For complete details, consult *49 CFR 192.707*.

CALL INVESTIGATION REPORT

COMPANY NAME:	**DATE:**
CALLER'S NAME:	**DATE:**
CONTACT TELEPHONE NUMBERS/NAMES:	
NAME:	**NUMBER:**
NAME:	**NUMBER:**
AREA/LOCATION OF REPORT:	
DESCRIPTION:	
NOTIFIED NAME:	**DATE:**
CONTROLLER'S NAME:	**TIME:**

103F02.EPS

Figure 2 ◆ Call investigation report sheet.

Line markers must be designed so that they can be clearly seen and give enough information to the public. Line markers must include the following information:

- Clear lettering on a sharply contrasting background
- The words *WARNING, CAUTION,* or *DANGER* followed by the words *Natural Gas Pipeline*
- Letters that are at least 1-inch high and ¼-inch thick
- The name and telephone number of the company operating the line

3.2.4 Audits

As you have learned, the federal government requires pipeline operators to have public education programs. Therefore, to show that they are complying with the law, operators may do some or all of the following tasks:

- Keep records on, or audit, the programs
- Assign staff to review programs to ensure that they meet government requirements and industry standards
- Survey customers for feedback about educational materials

Most pipeline operators keep copies on file of all educational materials, press releases, news stories about public meetings, and other materials that demonstrate their compliance with the law.

3.3.0 Handling Emergency Reports

Emergency reports must be handled quickly and safely. Sometimes the person who calls with an emergency report may be upset. Thus, it may be difficult to get clear, accurate information before the caller hangs up. Therefore, it is important to quickly get as much information as possible. Staff members who take these calls should follow these guidelines:

- Stay calm
- Get the caller's name and telephone number
- Ask for an exact location of the release
- Ask for directions or landmarks that can guide response teams to the area
- Instruct the caller to stay a safe distance from the area
- Quickly alert the appropriate person or persons in the company

Some emergencies involve local emergency response agencies. However, calling in outside help may not always be necessary. Often the pipeline company can handle the problem safely without calling for outside help and without creating alarm. In addition, it is important to remember that in some cases, an emergency report may be a hoax. Therefore, many operators require employees who are authorized to deal with emergency reports to do the following tasks:

- Investigate and evaluate the situation
- Carry out established company safety procedures to protect people, property, and the environment
- Alert the appropriate person or persons in the company
- Alert and work with emergency response agencies only when directed to do so by authorized supervisors
- Notify the appropriate regulatory agencies if required to do so
- Complete follow-up reports as required by law and company policy

NOTE

Damage to the pipeline may be the result of a criminal or terrorist act. In these cases, law enforcement agencies must be notified. This notification should come from someone in the company authorized to make such contacts.

3.4.0 Communicating with Emergency Response Organizations

Counties in every state are required by law to set up and maintain a local emergency planning committee (LEPC). Membership in these committees includes people from state and local government, law enforcement, civil defense, firefighting companies, environmental and transportation agencies, hospitals, the media, community groups, and local industries. LEPCs gather information on possible hazards in the community. They also draw up response plans for emergencies.

Pipeline operators must meet with LEPCs and the local emergency management director on a regular basis. Following are examples of some of the activities that a pipeline operator may participate in:

- Attending informational meetings where committee members learn about available emergency resources
- Scheduling visits to the pipeline facility by emergency response agencies
- Providing maps and blueprints to help emergency response agencies learn about the facility and plan what to do in an emergency

- Making plans for how to respond to emergencies and how to communicate when an emergency occurs

3.5.0 Communicating with the National Response Center

For some types of incidents, an authorized company employee must call the National Response Center (NRC). The NRC is a regulatory agency responsible for taking reports about emergency spills or leaks from oil and gas pipelines. A federal law describes what types of incidents to report and how to report them (*49 CFR 191.3*). This law applies to natural gas, flammable gas, toxic gas, and corrosive gas. Incidents that must be reported to the NRC include these events:

- A release of gas from a pipeline
- A release of liquefied natural gas
- Death
- Hospitalization
- Property damage, including the cost of gas lost, of $50,000 or more
- Emergency shutdown
- Any emergency that the pipeline operator judges important enough to report to the NRC

When one of these incidents happens, the authorized employee must call the NRC as soon as possible and give the following information:

- Name and telephone number
- Name and telephone number of the pipeline operator
- A description of what happened
- Location and time of the emergency
- Number of deaths and injuries
- Amount of damage

It may be possible to give only estimates in a first report. The pipeline company is responsible for making follow-up reports so that the correct information is on file with the NRC.

3.6.0 Communicating with Federal, State, and Local Agencies

As you have seen, the government regulates many pipeline operations. Therefore, calls from federal, state, and local government agencies are common. Such calls usually concern legal matters and can involve complicated information. Staff members responsible for answering these calls should obtain this information:

- The caller's name and position or title
- The name of the government agency
- A brief description of the purpose of the call
- The caller's telephone number and times when the caller can be reached

The person who must call the government agency back should find out the reason for the call before returning it. That person may have to do some research or contact others in the company to gather information. Knowing the purpose of the call helps to save time.

3.7.0 Communicating with the Media

Only authorized company personnel may speak with the media. Some pipeline operators may have an official spokesperson. This person handles media requests for information, press releases, and press conferences. Pipeline staff authorized to speak with the media must ensure that their facts are correct before making a statement. Inaccurate statements might damage the company's reputation or even cause panic.

If a reporter cannot be put in touch with the appropriate person immediately, the pipeline staff responsible for answering the telephone must do the following tasks:

- Write down the caller's name, organization, and telephone number
- Ask for a time that the caller can be reached, and write down a brief description of the caller's request
- Assure the caller that the message will be given to the appropriate person in the company as soon as possible
- Deliver the message to the appropriate person

Reporters often work on a deadline. Therefore, they might try to get information right away from the person who has answered the phone. If a reporter asks for information or even just an opinion, it is best to state that only authorized personnel may speak with the media.

3.8.0 Communicating with Construction Companies

Usually construction companies use the One-Call notification system to alert the pipeline company of a planned excavation near the pipeline. However, they may contact the company directly. The staff person taking the call from a construction company must obtain this information:

- The name and telephone number of the person responsible for overseeing construction (usually the general contractor)

- The name of the construction company
- A description of the work planned
- A complete description of where the planned construction will take place, which may be an address or geographical location
- Any other information to help pinpoint the exact location of the planned excavation

Pipeline personnel are then assigned to monitor the construction work. On more complex projects, pipeline personnel and the contractor have **daily tailgate meetings.** These meetings are held at the construction site often with a truck's tailgate serving as a conference table. The contractor and pipeline personnel discuss the day's excavation work and anything that might affect the pipeline's safety, including weather, the use of explosives, and the use of heavy construction equipment. In addition, pipeline personnel perform these tasks:

- Inform the contractor about the company's safety and excavation policies
- Give instructions to protect the pipeline from damage
- Decide how many inspections are needed to protect the pipeline
- Decide what types of inspections are needed
- Deal with any problems
- Submit required reports

4.0.0 ◆ ONE-CALL NOTIFICATION SYSTEM

Pipes used on the pipeline are strong, and they are carefully installed to ensure safety. However, if the earth around a pipeline is disturbed, the line may be damaged. Damage to a line can result in injuries, death, and destruction of property. Anyone planning to dig, plow, or excavate needs a simple and sure way to find out if the work is located near a pipeline.

To protect the public and the pipeline, the federal government requires all states to set up and maintain a One-Call notification system. The system covers both large transmission lines and smaller distribution lines. The One-Call notification system proceeds as follows:

- Anyone planning to dig, plow, or excavate must contact the local One-Call system
- The One-Call operator notifies the pipeline operator
- The pipeline operator clearly marks the pipeline's location

The government has strong penalties for anyone who does not use the One-Call notification system. Excavators who do not use this system or who ignore existing pipeline markers may be fined, sent to prison for five years, or both.

4.1.0 Excavator's Responsibility

Generally, excavators must call the One-Call system at least 48 hours before digging, but the lead time may vary from state to state. In an emergency, excavators are allowed to begin digging immediately. However, they must still contact the One-Call center as soon as possible. In either case, the excavator must give complete details, including this information:

- The company name, address, and telephone number
- The name of the person responsible for overseeing the excavation
- The exact location of the work
- The date the work will begin
- A description of the work

4.2.0 One-Call's Responsibility

Employees staffing the One-Call system must have accurate and up-to-date maps and contact information for pipeline operators in their area. These tools allow them to find the exact location where work is planned and identify the pipeline system or systems affected by the work. One-Call system employees must promptly call the pipeline operator or operators affected by the planned work and provide all the details they collected from the excavator.

4.3.0 Pipeline Operator's Responsibility

Federal laws outline the responsibilities of companies that operate underground pipeline systems (*49 CFR 192.614* and *49 CFR 195.442*). These responsibilities can be divided into two broad areas: the operator's responsibility to the public located near the pipeline and the operator's responsibility to the One-Call system.

4.3.1 Responsibility to the Public

As discussed earlier in this module, federal law requires pipeline operators to have a damage prevention program (*49 CFR 192.614*). That program must include a plan for alerting everyone in the vicinity of the pipeline in case of an emergency.

The term *vicinity* is not defined in the law. Generally, the company decides how large the area should be. This decision is made with help from the appropriate regulatory officials. To fulfill its responsibility, the pipeline company maintains the following lists:

- The names, addresses, telephone numbers, and map locations of people and businesses in the vicinity of the pipeline
- The names, addresses, and telephone numbers of companies who do excavation work in the vicinity of the pipeline

These lists must be kept up to date, and personal visits are the best way to regularly update the information on file. Company representatives visit each home and business in the defined vicinity to confirm information. During a personal visit, a company representative can complete the following tasks:

- Add addresses for new buildings or make note of new construction
- Remove addresses for buildings that no longer exist
- Confirm location information for people who may not have a telephone

Many operators update their lists with personal visits at least every 12 to 15 months. These visits also provide an opportunity to give out information about the One-Call system.

When producing information about the One-Call system, most pipeline operators use a standard logo that is easy to recognize (see *Figure 3*). They may also give away items, such as stickers or pencils, that have One-Call information printed on them.

Dig Safely.
1-888-258-0808

103F03.EPS

Figure 3 ◆ Dig safely logo.

4.3.2 Responsibility to the One-Call System

A pipeline company's management is responsible for making sure that the One-Call system guidelines are followed. Operators generally list employee responsibilities in their training manuals. A training manual may include the following information:

- How to handle One-Call notifications
- How to document the call and any actions taken by the company
- How to complete and file necessary reports

The staff member who takes the One-Call notification must promptly notify the appropriate supervisor. The supervisor then assigns personnel to mark the pipeline and, if necessary, monitor the work.

Summary

You quickly write down a phone message for directions to what sounds like Wyatt Street, but what the caller actually said was Wyant Street. Your boss asks you to buy 25 pizzas, but you never get the message that changes the order to 10. Your neighbor plans to build a deck the same day you plan to have an outdoor party. Because neither of you mentions these plans, both of you are embarrassed and annoyed. These are examples of poor communication. No matter whose fault it is, the result is frustration or a waste of time or money.

On the pipeline, errors in communication can have a much bigger price tag. Errors can result in damage to expensive equipment, injuries, or even death. In fact, communication in this industry is so important that it is covered by several federal laws. This module provided an overview of effective communication in the pipeline industry. You must always be aware that other people—co-workers and the public—are affected by the work you do. Whether you are giving instructions to a co-worker, sending in a report from a patrol, or writing down information about an emergency, good communication skills help you do your job better and more safely.

Review Questions

1. The control center records operating changes on the _____.
   a. SCADA inventory form
   b. operations control form
   c. daily operating sheet
   d. work order

2. A(n) _____ is issued when pipeline workers must report to work outside their normal work hours.
   a. call-out
   b. emergency response call
   c. special work order
   d. overtime authorization order

3. A request from a buyer to move gas through a pipeline to a delivery point is called a(n)_____.
   a. online purchase order
   b. gas market order
   c. nomination
   d. election

4. Line markers must _____.
   a. be bright orange
   b. be no more than 20 feet apart
   c. have letters at least 1-inch high and ¼-inch thick
   d. have signal flags attached when placed near railroad crossings

5. Pipeline operators audit educational programs to _____.
   a. account for the cost of the programs
   b. avoid repeating programs already taught
   c. improve quality
   d. demonstrate compliance with federal law

6. To effectively plan for emergencies, authorized pipeline personnel must meet regularly with _____.
   a. the fire department and a hazardous materials committee
   b. the city management director and the mayor
   c. appointed government officials and the police department
   d. the local emergency planning committee and the local emergency management director

7. When an incident results in hospitalization, death, or property damage of $50,000 or more, the _____ must be notified.
   a. DOT
   b. NRC
   c. LEPC
   d. CFR

8. Only _____ may speak with the media.
   a. the marketing department
   b. field supervisors
   c. the vice president of the company
   d. authorized personnel

9. Excavators who do not use the One-Call notification system or who ignore pipeline markers may _____.
   a. be fined, sent to prison for five years, or both
   b. be fined or sent to prison for ten years, or both
   c. have their contractor's license revoked
   d. have to pay a special fine to the pipeline company

10. When there is no emergency, excavators generally must notify the One-Call system at least _____ before digging.
    a. 12 hours
    b. 48 hours
    c. one week
    d. two weeks

GLOSSARY

Trade Terms Introduced in This Module

Call-out: A call for extra employees, initiated by a supervisor, to report to work to resolve an immediate or urgent problem.

Daily operating sheet: A record documenting all operating events of the day. It is sometimes called a log.

Daily tailgate meeting: A meeting attended by assigned pipeline personnel and the excavation or construction contractor to monitor conditions that might affect pipeline safety.

End users: The customers who receive and use natural gas for heating, cooking, or industrial processes.

Field: (1) Any outdoor area where the pipeline is located. (2) The area where natural gas is processed and refined. This term may also be used to refer to employees who work in the field.

Line pack: A term that describes the storage of natural gas in transmission lines.

Marketer: A company or group of companies who purchase gas supplies for resale to utilities or end users.

Nomination process: The method used to request the transmission of gas through a pipeline to a delivery point. The nomination system allows customers to reserve time in the pipeline for delivery.

One-Call notification system: A service that helps protect buried pipelines and public safety. Anyone planning to dig can call a toll-free number. The service then contacts pipeline operators so that they can mark the pipeline location.

Operations: (1) All of the processes involved in transporting natural gas through the pipeline to the appropriate delivery points. (2) Field and control center employees who are responsible for moving natural gas through the pipeline to the appropriate delivery points.

Operator: The legal entity that owns or operates the pipeline facility.

Right-of-way (ROW): The area along the pipeline that the pipeline is legally permitted to use to construct, operate, and maintain the pipeline. The ROW extends the entire length of the pipeline. Rights of way can be 50 to 100 feet wide during construction, but the ROW typically becomes much smaller after construction is completed. It may be no wider than just a few feet on either side of the pipeline itself, or it may even be just the width of the pipeline.

Supervisory Control and Data Acquisition (SCADA): A computer system that is used to monitor and control the operation of a pipeline.

Work order: A written authorization that describes what work is to be performed and when and where it is to be performed.

Figure Credits

Koch Pipeline Company, LP	103F01
Office of Pipeline Safety	103F02

CONTREN™ LEARNING SERIES — USER UPDATES

The NCCER makes every effort to keep these textbooks up-to-date and free of technical errors. We appreciate your help in this process. If you have an idea for improving this textbook, or if you find an error, a typographical mistake, or an inaccuracy in NCCER's Contren™ textbooks, please write us, using this form or a photocopy. Be sure to include the exact module number, page number, a detailed description, and the correction, if applicable. Your input will be brought to the attention of the Technical Review Committee. Thank you for your assistance.

Instructors – If you found that additional materials were necessary in order to teach this module effectively, please let us know so that we may include them in the Equipment/Materials list in the Instructor's Guide.

Write: Curriculum Revision and Development Department
National Center for Construction Education and Research
P.O. Box 141104, Gainesville, FL 32614-1104

Fax: 352-334-0932

E-mail: curriculum@nccer.org

Craft ______ Module Name ______

Copyright Date ______ Module Number ______ Page Number(s) ______

Description ______

______

______

______

(Optional) Correction ______

______

______

(Optional) Your Name and Address ______

______

______

Module 67104-02

Routine Field and Facility Operations
(CT 43, 50, 51, 54, 56, 57, and 58)

Introduction to the Pipeline Operator Qualification Rule, API, and NCCER

It could be said that petroleum products keep America running. Petroleum products like oil and natural gas are the principal sources of energy in the United States and are used in many manufacturing and chemical processes. Petroleum products are primarily transported by pipeline. There are more than a million miles of liquid and natural gas pipeline in this country, ranging in diameter from two inches to four feet. These pipelines bring crude oil and natural gas from production fields to processing plants and refineries—and ultimately to the consumer.

Pipeline Operator Qualification

Petroleum products are considered hazardous, so pipeline personnel must take extra precautions and follow specific procedures when working with them. To help ensure the safe handling and transportation of petroleum products, the United States Department of Transportation (DOT) mandates safe practices as federal regulations.

One such regulation, the Operator Qualification (OQ) rule, requires pipeline operators to develop a written qualification plan to evaluate their personnel and their contract personnel (see *49 CFR Part 192 Subpart N—Gas*, and *49 CFR Part 195 Subpart G—Liquid*). The evaluation hinges on the operator's ability to perform certain covered tasks and to be able to recognize and respond to abnormal operating conditions that may be encountered while performing these covered tasks.

Covered Tasks

A pipeline performance task is deemed *covered* if the activity passes the following four-part test:

1. Is it performed on a pipeline facility?
2. Is it an operations or maintenance task?
3. Is it performed as a requirement of *49 CFR Part 195 (Liquid)* or *49 CFR Part 192 (Gas)*?
4. Does it affect the operation or integrity of the pipeline facility?

If the answer to all four questions is *yes*, then the activity is considered *covered* and must figure into the evaluation process of pipeline personnel.

American Petroleum Institute

The OQ rule and its related covered tasks provided the catalyst to develop a unified pipeline training and assessment program—so that pipeline operators and owners have ready access to the assessment and training tools they need. This effort was the direct result of the vision and leadership provided by The American Petroleum Institute (API) and API members.

Efforts to unify the industry are nothing new to API. Indeed, the impetus for forming the Institute in 1919 was the need to standardize engineering specifications for drilling and production equipment. Today, the most pressing issues revolve around public perceptions and government policies toward the industry. Speaking with one voice on these issues has become as essential as having interchangeable parts in the field. API's members determine the petroleum industry's positions on public policy as well as the standards that govern its day-to-day operations worldwide.

Consortium on Operator Qualification

API assembled a focus group whose primary purpose was to identify covered tasks and provide compliance recommendations to pipeline operators.

The Consortium on Operator Qualification (COOQ) became the guiding force for the identification and development of covered tasks for the liquid pipeline industry. Their work product, API publication 1161, *Guidance Document for the Qualification of Liquid Pipeline Personnel*, formed the basis for the resulting Pipeliner Training and Assessment Program.

The covered task list created in API publication 1161 has now evolved to include gas covered tasks and is maintained by NCCER. The most current covered task list can be downloaded at www.nccer.org.

National Center for Construction Education and Research

With the guidance and support of API, its members, and the Consortium on Operator Qualification, the National Center for Construction Education and Research (NCCER) was able to design and implement the Pipeliner Training and Assessment Program.

NCCER is a not-for-profit 501(c)(3) education foundation established in 1995 by the world's largest and most progressive construction and maintenance contractors and national contractor associations. NCCER was founded to address the

severe workforce shortage facing the construction and maintenance industries and to develop a standardized training process and curricula. Today, NCCER is supported by hundreds of leading construction and maintenance companies, manufacturers, and national associations and developed more than 35 construction and maintenance titles.

NCCER also maintains a National Registry, which contains the records of trainees and instructors who have received NCCER training. All completion information for every trainee is sent to the NCCER and filed within the Registry. The Registry can then confirm training and skills for workers as they move from company to company, state to state, or even within their own company.

Standardized Curriculum (NCCER's Contren™ Learning Series)

NCCER curricula are developed using the expertise of industry subject matter experts (SMEs). The use of SMEs as an integral part of every new development or revision ensures that the curriculum is current, industry-driven, and indicative of industry practices. The development and/or revision process is facilitated from start to finish by NCCER project managers, who work closely with the SMEs and a select team of technical writer(s) on the development and presentation of the curriculum.

Pipeline: Five Training Paths

The pipeline curriculum focuses on essential skills and competencies within five skill areas:

- Corrosion Control
- Electrical and Instrumentation
- Maintenance
- Mechanical
- Operations (Field and Control Center)

SMEs from the pipeline industry have determined the skills and competencies needed based on API's covered tasks and the consensus of the group/industry. Specific company information is generally not incorporated into the curriculum unless the SMEs determine that the information is the industry standard.

Training Modules and Covered Tasks

The smallest unit of instruction in the curriculum is a *module*. Each curriculum is composed of modules (the equivalent of stand-alone chapters) that are logically sequenced into complete *levels* of instruction. Every module is task-oriented and requires the successful completion of a written (knowledge-based) examination and a performance (hands-on) test. Modules may be individually selected and/or combined for efficient task training. Modules in the pipeline curricula, to the extent possible, will align with API-developed covered tasks. While not all modules will relate to covered tasks, all covered tasks will relate to at least one module.

Benefits of Training to the Pipeline Marketplace

Applying NCCER's standardized training process allows pipeline owners and contractors in the United States to establish the benchmark for acceptable skill and competency levels within the pipeline industry. This process will assist training organizations in developing and/or improving standardized pipeline training programs that effectively match their resources and operating conditions while responding to regulatory demands and to the needs of the industry. The process will also recognize training organizations that meet the minimum standards of competence, warranting regulatory, public, and professional confidence. This in turn will foster national unity within the pipeline industry for the mutual benefit of individual technicians, owners, contractors, and ultimately pipeline-users and society.

Additional Resources

To review the Pipeliner Training and Assessment Program Implementation and Resource Guide and the latest version of pipeline covered tasks, visit NCCER's Web site at www.nccer.org, or contact:

National Center for Construction Education and Research
PO Box 141104
Gainesville, FL 32614-1104
(888) 622-3720
Web site: www.nccer.org

For more information on API, please contact:

American Petroleum Institute
1220 L Street, Northwest
Washington, D.C. 20005
(202) 682-8000
Web site: www.api.org

To order *API Pub 1161*, contact Global Engineering Documents by calling 1-800-854-7149 or visiting their Web site at www.global.his.com.

Additional sources of information on the pipeline operator qualification regulation include:

DOT – Office of Pipeline Safety
www.ops.dot.gov

Transportation Safety Institute
www.tsi.dot.gov

COURSE MAP

This course map shows all of the modules in the Gas Pipeline Operations curriculum. The suggested training order begins at the bottom and proceeds up. Skill levels increase as you advance on the course map. The local Training Program Sponsor may adjust the training order.

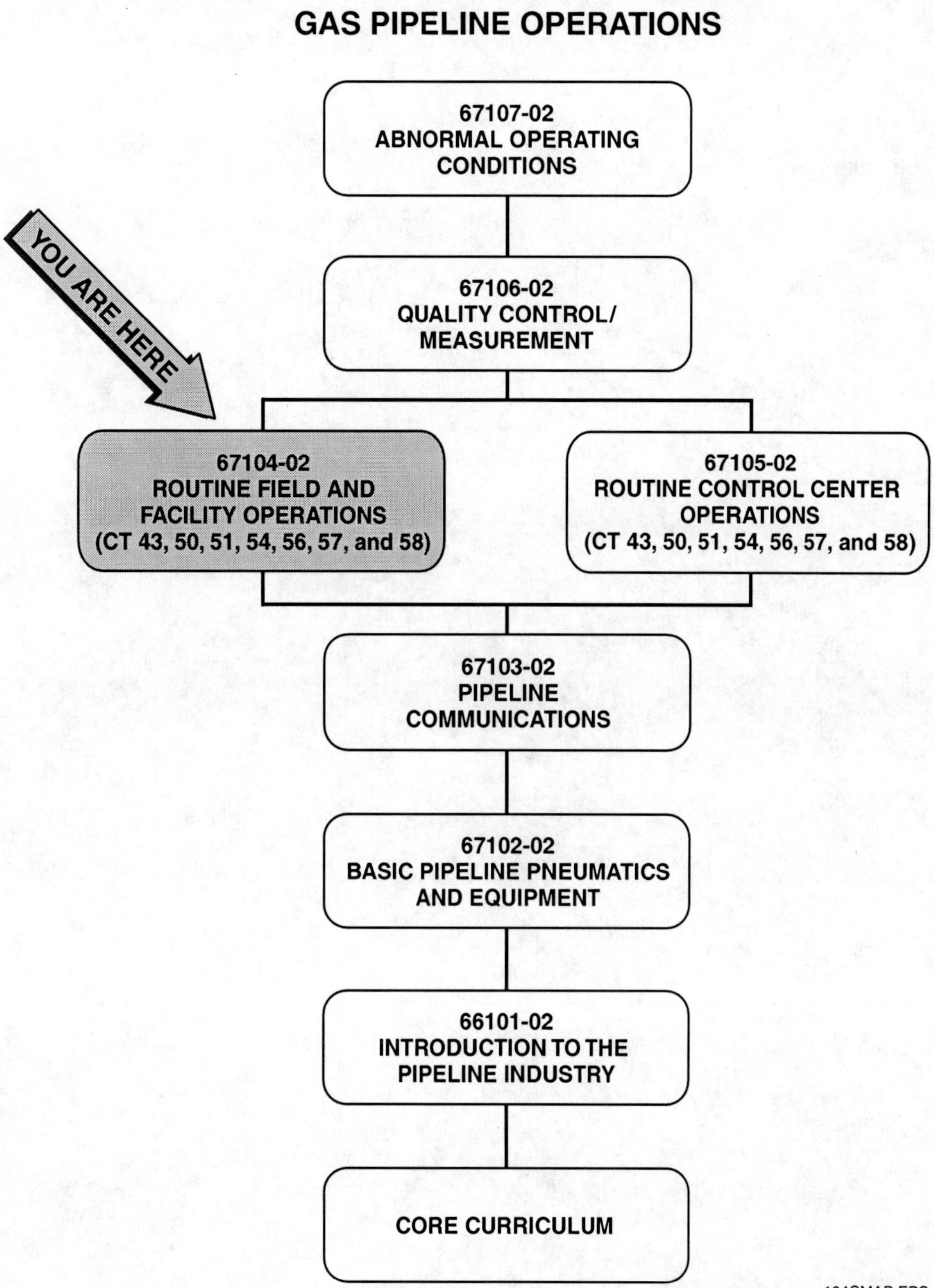

104CMAP.EPS

MODULE 67104-02 CONTENTS

Figures

Table

MODULE 67104-02

Routine Field and Facility Operations (CT 43, 50, 51, 54, 56, 57, and 58)

Objectives

When you have completed this module, you will be able to do the following:

1. Perform routine facility inspections and monitor pipeline parameters (CT 43.3) to ensure normal facility operations.
2. Describe the general procedures for performing system startup (CT 43.1) and system shut in/shutdown (CT 43.2).
3. Operate valves (CT 43.4) and compressor units and monitor engines.
4. Test remotely controlled shutdown devices (RCSDs) (CT 54).
5. Monitor alarms and safety devices.
6. Identify, select, launch, track, and receive pigging devices.
7. Purge gas from pipelines (CT 50), purge air from pipelines (CT 51), and monitor purging during maintenance.
8. Monitor weather conditions.
9. Locate a buried pipeline.
10. Describe the general procedures involved in uprating the pipeline's maximum allowable operating pressure (MAOP) (CT 56).
11. Operate odorant equipment (CT 57) and monitor odorant level if applicable (CT 58).
12. Perform right-of-way (ROW) inspections, linewalking, and surveillance.

Prerequisites

Before you begin this module, it is recommended that you successfully complete the following: Core Curriculum; Pipeline Core, Module 66101-02; Gas Pipeline Operations, Modules 67102-02 and 67103-02.

Required Trainee Materials

1. Pencil and paper
2. Appropriate personal protective equipment

1.0.0 ◆ INTRODUCTION

Personnel involved in routine field and facility operations must know how to properly operate, monitor, and maintain pipeline equipment. In addition, they must know when conditions are normal so that they can quickly identify any conditions that are not normal. Any operating condition that is not normal is classified as an **abnormal operating condition (AOC).** Knowing how to identify and properly respond to AOCs is an important skill for pipeliners. Many AOCs are handled routinely, but some may result in an emergency situation. In emergencies, pipeliners must follow their companies' established safety procedures to protect people, property, and the environment.

As part of their daily routine, field personnel, working under the direction of supervisors, receive information from and provide information to the control center. This process helps to ensure that anything that occurs on one part of the pipeline, such as repair work, does not adversely affect other operations. This module presents the basic responsibilities of field personnel. It describes the areas in which field personnel must develop expertise in order to keep pipeline operations safe and efficient. In addition, it describes functions that field workers should understand even though they may not perform the functions in their jobs.

2.0.0 ◆ ROUTINE FACILITY INSPECTIONS

In the field, pipelines and other facilities must always be checked carefully. Pipeline companies use computer technology to keep track of all of their operations from the control center, but nothing beats the eye of an experienced employee when it comes to spotting many problems. Sometimes an employee has to go into the field to figure out why equipment is malfunctioning.

Pipeline companies and the U.S. Department of Transportation (DOT) require regular inspections of facilities to ensure that they are kept in good condition. In the field, the station operator usually performs inspections and possibly makes repairs. Routine inspections cover the following items:

- General appearance
- Mechanical problems
- Physical condition of the property

2.1.0 General Appearance

Personnel ask several questions when checking the general appearance of a facility:

- Does the facility look clean overall?
- Is there evidence of vandalism or trespassing?
- Are the fences, gates, locks, no-trespassing signs, and other security measures in good condition?
- Has nature caused any problems? For example, has there been any damage from high water or strong winds?

Inspecting a facility also requires using senses other than sight. When considering sounds and smells, personnel ask the following questions:

- Are there any loud mechanical noises from equipment?
- Is equipment running too hot?
- Are there strong odors?

2.2.0 Mechanical Problems

Valves, compressors, **drivers** (reciprocating engines and turbines), pumps, and other auxiliary equipment must all be checked carefully for problems. For example, a valve may fail to respond to a signal from the control center, a compressor may fail, or a driver may exhibit excessive vibration. It is especially important to be aware of conditions that suggest gas may be leaking.

Ordinarily, field personnel report problems, such as damage to the pipeline or auxiliary equipment, to the control center. Some companies, however, may have a different reporting procedure. Always follow company policy when reporting problems. Personnel must not risk injury by trying to solve problems that require special equipment or skills.

Other equipment in the field must also be checked carefully. This equipment includes the alarms as well as the devices that measure the pressure in the pipeline. Such inspections must be carefully documented.

2.3.0 Physical Condition of the Property

Station buildings must be inspected to ensure that they are clean and in good repair. Following are items that are commonly inspected:

- Doors and windows and their locks
- Roofs
- Railings
- Stairs
- Floors
- Lights
- Storage areas

The buildings must be clean, and parts, tools, and equipment must be carefully stored. Stairs and exits must not be blocked. Locks must be in good working order. Interior and exterior lights should function. Field personnel must report any condition that could affect safety or security.

2.4.0 Monitoring Pipeline Parameters

In order to effectively monitor a pipeline, you must be familiar with the normal operating pressure, flow, product characteristics, and other parameters for the pipeline. With this knowledge, you must continually monitor the pressure and flow, looking for any changes.

Overall pipeline parameters are monitored by the control center. When an alarm indicating a problem sounds, control center personnel notify field personnel. In most cases, field personnel investigate and evaluate the problem and report back to the control center. On the basis of this information, the field maintenance supervisor notifies all affected personnel and schedules repairs. The following conditions are monitored:

- Low pressure
- No pressure
- High pressure
- Hazardous atmosphere
- Equipment failure

2.5.0 Responding to Abnormal Operating Conditions

Pipeline personnel must respond to AOCs, such as a gas leak, equipment malfunction, or facility damage, quickly but not hastily. It is important to follow established company procedures so that each problem is confirmed, investigated, and addressed safely and correctly. Personnel confirm alarms to ensure that an alarm is not a false alarm. They investigate problems to determine what is wrong and what needs to be done to correct them. Depending on the situation, steps may have to be taken immediately to secure the area and ensure the safety of personnel. Other situations may require that repairs be scheduled and necessary work permits issued. In general, personnel follow these steps when responding to an AOC:

Step 1 Evaluate the situation.

Step 2 Notify affected personnel.

Step 3 Carry out company safety procedures to protect personnel, property, and the environment.

Step 4 Make necessary repairs or report that repairs are required.

Step 5 Submit the required reports.

In some cases a problem may develop into an emergency situation that may affect the public located near the pipeline. Companies have specific procedures for contacting outside assistance such as local emergency response agencies. Only authorized company personnel may call for outside assistance.

3.0.0 ◆ PERFORMING SYSTEM STARTUP AND SEGMENT SHUT IN

In gas pipeline operations a full system startup is performed only on a newly installed system. In addition, the entire pipeline system is almost never shut down. Usually, a **pipeline segment** is shut in, or isolated, for maintenance or repairs. Shutting in the segment allows the gas to keep flowing while maintenance or repairs are performed. Personnel should be familiar with the processes involved in starting up a system and shutting in a pipeline segment. The general steps involved in each process are covered in this section.

Personnel who perform a system startup or shutdown must possess basic knowledge in the following areas:

- Pipeline pneumatics
- The maximum allowable operating pressure (MAOP) for the pipeline system
- Company operations policies
- Company policies for the appropriate response to problems or AOCs

In addition, personnel must be familiar with the pipeline's normal operations. They must know what is normal so that they can quickly spot anything that is not normal.

3.1.0 Startup

There must be a clear flow path from the origin point to the delivery point before the pipeline is started up. All equipment must be functioning properly and ready to receive the gas flow, which moves downstream from the origin point. Compressors at downstream stations are activated by a pressure rise from upstream stations.

Specific steps for a startup procedure and the order in which the steps are taken vary from one company to another. In general, a typical startup may include some or all of the following steps:

Step 1 Confirm that the proper work orders and work schedules have been issued.

Step 2 Confirm that the gas is ready to enter the pipeline.

Step 3 Confirm that upstream and downstream stations are ready for startup. This step includes confirming that all necessary equipment is available.

Step 4 Confirm that all main-line station valves are in the normal position. Confirm that the flow path is correct.

Step 5 Start the compressors.

Step 6 Notify the next station downstream that the compressors are running.

Step 7 Start the compressor units at stations on the line that will be affected immediately.

Step 8 Notify the next station downstream that the compressor unit at that station is coming online.

Step 9 Set the compressor unit to the desired pressure. Watch for the rise in pressures downstream.

Step 10 Notify each station downstream that the compressor unit is coming online.

Step 11 Monitor the system until the pipeline is operating in a **steady state.**

Step 12 Ensure that all required documents have been completed.

3.2.0 Shut In

Sometimes a pipeline section must be shut in, or isolated, for inspection, maintenance, or repair work. This work can be performed without stopping the flow of gas by installing a **hot tap.** Hot tapping requires a **hot work** permit, and personnel are not allowed to perform this task until all the correct permits are obtained. The hot tap process allows personnel to install a loop of pipeline so that the gas can be diverted around the shut-in section. This loop is welded to the main pipeline with valve fittings at either end.

When a hot tap is installed, a fitting is welded on the main pipeline near the section to be shut in. A valve and tapping device are installed on this fitting, and the valve is opened. A pilot drill attached to the tapping device drills a hole in the pipeline. Gas moves from the pipeline into the tapping device, air is purged, and the valve is closed. Any remaining gas in the shut-in section is vented to the atmosphere. The valves on the loop pipeline are opened, allowing the gas to take a detour around the shut-in section. The shut-in section is checked to ensure that all gas and air have been purged before any repairs or inspections are performed. Depending on the circumstances, the loop may be a permanent or a temporary installation. If the loop is temporary, the process is reversed to shut in the loop and return flow to the main line.

Specific steps for shutting in a pipeline segment and the order in which the steps are taken vary from one company to another. Additional steps are required for a full system shutdown (if one should ever be required) and each company has its own procedures for this process. In general, a typical shut-in procedure may include the following steps:

Step 1 Confirm that the proper work orders, work permits, and work schedules have been issued.

Step 2 Complete maintenance on valves if necessary.

Step 3 Conduct a leakage survey over the segment to be shut in.

Step 4 Confirm that affected personnel are prepared to shut in the pipeline segment.

Step 5 Install a hot tap on the section to be shut in.

Step 6 Install the loop section of pipeline and monitor pressure as gas flows into the loop.

Step 7 Notify appropriate personnel that the pipeline segment is shut in.

Step 8 Monitor pressures in sections adjacent to the shut-in section to ensure the MAOP is not exceeded.

Step 9 Complete the required maintenance, repair, or inspection of the shut-in section.

Step 10 Reverse the process to remove the loop section of pipeline and restore gas flow to the shut-in section.

Step 11 Submit the required reports.

It is important to monitor the area for leaks during the entire shut-in process. Field personnel must wear the appropriate personal protective equipment (PPE) while performing this task.

3.3.0 Monitoring

Conditions on a pipeline can change without warning. That is why it is important to monitor conditions when a segment is shut in. Field personnel rely on the control center to monitor pipeline pressure and temperature to ensure they remain within acceptable limits. They also rely on safety devices at the work site to alert them to problems.

If there is an unexplained change in pressure, the general procedure is to first determine the cause and then control the situation. The control center may make adjustments to bring the pressure back to acceptable limits or alert field personnel to problems they must address. Field personnel may be directed to check any compressors or valves operated before the pressure change to ensure they are still in the right position. Similar procedures are followed in response to any other problems that could affect the safety of field personnel or the integrity of the pipeline.

Personnel must ensure that all required repairs have been completed. They must also receive authorization from the maintenance supervisor before resuming operations.

4.0.0 ◆ PURGING

Sometimes a pipeline section must be shut in so that it may be **purged** for inspection or maintenance. Gas is routed around the shut-in section so that there is no interruption in service. When isolating a pipeline section, field personnel must **unpack** and purge the contents. After the pipeline is emptied of gas, explosive vapors may still be present. Personnel must then purge the line to remove all hydrocarbon gases. Then personnel must do a second purge. They must purge the air in the pipeline with natural gas before repressurizing and **packing** (*49 CFR 192.69*).

Flammable gas explodes or catches fire when it is mixed with a certain percentage of air. Typically, a mixture of gas and air must be 5% to 15% gas and 85% to 95% air before it explodes. A mixture of gas and air that does not have enough gas to explode is below the **lower explosive limit (LEL).** It is too lean to explode. A mixture with more than the maximum amount of gas is above the **upper explosive limit (UEL).** It is too rich to explode. Purging establishes a moderately fast flow in the line being purged. This rapid flow helps to ensure that where the air and gas meet, they mix as little as possible. Personnel must maintain a constant purge rate until all air or gas has been expelled.

> **NOTE**
>
> The term *explosive limits* refers to the upper and lower percentages of gas, in combination with air, that are explosive. Each gas has its unique range. Typical ranges are from 5 percent to 15 percent of gas mixed with air.

4.1.0 Purging Gas from a Pipeline

To purge a transmission pipeline segment, personnel open valves and use an inert gas to push out the flammable gas. The employee in charge of the purging procedure defines the appropriate process to be followed. All employees engaged in the process must wear the appropriate PPE. Before and during purging operations, these personnel must do the following tasks:

- Follow the company operations and maintenance (O&M) procedures and be aware of possible hazards along the pipeline ROW.
- Monitor for hazardous vapors throughout the purging process.
- Know and follow the company's lockout/tagout procedure and know how to obtain the appropriate work permits.

Following are the basic steps involved in purging gas from a pipeline:

Step 1 Safely shut in, vent, and disconnect the pipeline section that must be purged.

Step 2 Unpack and purge the segment.

Step 3 Purge the air in the pipeline with natural gas before repressurizing and packing the line.

Step 4 Reconnect all equipment after the purging is completed.

Step 5 Submit the required reports.

Local distribution companies may have additional procedures to purge the distribution lines that serve homes and businesses. Their procedures also include a process for quickly and safely restoring service.

4.2.0 Isolating the Section to Be Purged

Isolating the section to be purged requires following the company's lockout/tagout procedure. This procedure must be followed when hazardous materials could adversely affect maintenance. It includes isolating all such materials from the work area. The isolation must continue until the work is completed. Personnel must close off all valves and other devices that are directly connected to the section being shut in. These devices must not be accidentally opened while the purging operation is occurring.

Bonding provides a safe path for electric current across any gap in the pipeline while the purging is occurring. Sparking or arcing should be prevented during the purging process.

Personnel must measure the explosive level of hazardous vapors with a **combustible gas indicator (CGI).** No one should rely on their sense of smell to detect hazardous vapors for the following reasons:

- The concentration levels may not be high enough to detect dangerous levels by the sense of smell alone.
- A medical condition may affect a person's sense of smell.
- Certain hazardous vapors may deaden the sense of smell.

4.3.0 Purging Air from Pipelines

When there is air in a pipeline that could form a hazardous mixture with natural gas, that air must be purged from the pipeline. This process involves opening valves and using natural gas to push out the air. The procedures for purging air from a pipeline are the same as those for purging gas with the exception that you must sample the vent stream with a combustible gas indicator or portable gravitometer to determine when the vapor reaches 100% gas in air. Personnel must follow all of the same safety procedures.

4.4.0 Monitoring During Purging

During the purging process, personnel must monitor a number of conditions to ensure that the process is carried out safely. The following conditions are monitored:

- The air for hazardous gases
- The weather, especially wind direction
- All communications from the control center, especially those that alert personnel to hazardous conditions

When conducting a purging operation near a populated area, pipeline personnel must take special care to prevent hazardous vapors from drifting into that area. They must monitor vapors to ensure that they are not in the explosive range. Purging operations must be suspended if weather conditions do not allow the vapors to safely disperse.

5.0.0 ◆ OPERATING VALVES

Valves are mechanical units that perform two functions. They control the flow through a pipeline or open or shut off a pipeline completely. Some valves are automatic or semi-automatic safety devices.

There are many kinds of valves; among the most common are check, gate, plug, needle, and pressure-relief valves. Most valves have position indicators, which are shafts, rods, or pointers that show whether the valves are open or closed.

Most valves are operated from the control center. However, field personnel must know how to safely open or close valves manually. These operations must be performed carefully to safely control flow and pressure and maintain normal operations.

5.1.0 Valve Placement and Alignment

Valves are put at specific points on pipelines for specific reasons and they must be opened and closed in the right order. Opening and closing valves in the right order is called **valve alignment.** Companies have specific valve alignments for every operation. These alignments are laid out for employees on charts or lists so that errors can be avoided. Errors can be very costly. If valves open or close unexpectedly, the following problems can occur:

- Delivery and billing problems
- Lost time and work
- Ruptures in the pipeline

Many valves are operated electrically by remote control. Multiple valves are sometimes grouped into a **manifold,** also known as a header. Backup generator units or battery-operated units can be used to control the valves during a power failure. These valves can also be operated manually. Some valves, however, must be operated only by hand.

Field personnel must know where the valves in their work area are located and how to operate them manually if necessary. Manual operation may be required when valves do not respond to a signal from the control center. In addition, field personnel must know how to repair malfunctioning valves.

5.1.1 Failure to Operate

An alarm in the control center may signal that a valve is not operating. In addition, operations personnel may notice that a valve indicator is not moving up and down as it should. When there is a problem with a valve, field personnel may be asked to investigate and evaluate the situation. They report their findings to the maintenance supervisor or the control center, and a decision about how to address the problem is then made.

5.1.2 Leaking Valves

If a valve is leaking, it is not safe to manually operate it until the leak is stopped and the air cleared. Hazardous distillate or gas vapors can build up around a leak and catch fire or explode. Personnel must isolate the leak if it is safe to do so. The air must be monitored with a CGI. When hazardous vapors have been cleared, repairs can be made. The air must be continuously monitored during repairs. Once repairs have been made, personnel must submit any required reports.

Review Questions

Sections 1.0.0–5.0.0

1. When you are inspecting a facility, all of the following should alert you to a possible problem except _____.
   a. loud mechanical noises
   b. strong odors
   c. segment isolation
   d. broken lights

2. Ordinarily, field personnel report problems, such as damage to the pipeline or auxiliary equipment, to _____.
   a. local emergency response agencies
   b. company headquarters
   c. the engineering supervisor
   d. the control center

3. The _____ process allows personnel to install a loop of pipeline so that the gas can be diverted around the shut-in section.
   a. hot tap
   b. cold tap
   c. compression
   d. detour

4. A mixture of gas and air that does not have enough gas to catch fire is too _____ to explode.
   a. rich
   b. odorized
   c. compressed
   d. lean

5. Opening and closing valves in the right order is called _____.
   a. consignment
   b. alignment
   c. refinement
   d. assignment

6.0.0 ◆ OPERATING COMPRESSORS

Natural gas is sent through a pipeline at high pressures. As the gas moves through the pipeline, friction reduces its speed and pressure drops. To maintain the necessary pressure, the gas must be recompressed periodically as it moves through the pipeline. This is why **compressor stations** are installed approximately every 100 miles along the pipeline. Often, several compressors are located at one station.

Some compressors use a small part of the gas that flows through the pipeline to power the engine that drives the compressor. Some compressors are electrically powered by turbine engines. An example of a field compressor is shown in *Figure 1.* A natural gas compressor is shown in *Figure 2.*

Several different types of compressors are available for use on the gas pipeline. Therefore, it is important to read and follow the manufacturer's recommendations for installation, startup, shutdown, and maintenance procedures. These procedures should be available at the pipeline facility for easy reference if needed.

104F04.EPS

Figure 1 ◆ Field compressor.

104F02.EPS

Figure 2 ◆ Natural gas compressor.

6.1.0 Compressor Startup

The procedures for starting up a compressor and the order to follow vary from one company to another. In addition, each company has its own notification and reporting requirements. In general, the following steps are taken to start a compressor:

Step 1 Carry out any pre-startup procedures as recommended by the manufacturer or required by company policy.

Step 2 Review and verify knowledge of the starting unit and startup procedures.

Step 3 Ensure that the compressor and station valves are properly positioned.

Step 4 Notify all affected personnel that the compressor is about to be started.

Step 5 Confirm that all operating variables are within the safe operating limits of the compressor.

Step 6 Confirm that all safety alarms and shutdown devices are set and are working properly.

Step 7 Initiate the startup according to the posted startup procedure for the compressor(s) at your location.

Step 8 Follow the manufacturer's instructions for operating the compressor after it has been started.

The manufacturer's instructions include information regarding monitoring. The following items are monitored during operation of the compressor:

- Suction and discharge pressures
- Compressor operating temperatures, pressures, and speeds
- Unusual noises or vibrations
- Leaks
- Surges

6.2.0 Compressor Shutdown

Compressors and drivers may have to be shut down for regular maintenance or in response to a problem. This procedure must be performed carefully, safely, and in accordance with the manufacturer's specifications. In general, the following steps should be taken to shut down a compressor:

Step 1 Slow down the compressor to minimum speed.

Step 2 Unload the compressor and open the bypass valve.

Step 3 Allow the compressor to cool down.

Step 4 Shut off the fuel supply.

Step 5 Close the suction and discharge valves.

Step 6 **Blowdown** the compressor.

The amount of time that the compressor may be left depressurized depends on a number of factors, including the condition of the compressor, the need to make repairs, and the availability of backup compressors. Generally, the maintenance supervisor decides how long the compressor should remain out of service.

6.3.0 Abnormal Operating Conditions

Several AOCs are associated with startup and shutdown of a compressor. When AOCs occur, personnel must follow company policy for identifying the problem and notifying the appropriate personnel. In addition, personnel must conduct any investigations or submit reports required by law or company policy. Following are two common AOCs:

- *Gas leak*—A hazardous gas alarm will sound. Personnel may be required to leave the area. The gas leak must be repaired and the air checked with a CGI to ensure safety.
- *Excessive vibration or noise*—Too much vibration or noise may indicate that equipment is malfunctioning or about to fail. The compressor may have to be shut down while personnel investigate the cause.

7.0.0 ◆ MONITORING ENGINES

Engines that operate pipeline equipment such as compressors are in constant motion, which results in a lot of wear and tear on the parts. Engines must be monitored periodically to ensure that equipment does not fail. Two common problems that occur on engines are excess vibration and driver surge.

7.1.0 Excess Vibration

Every engine vibrates as it operates. Too much vibration, however, can damage it. Causes of harmful vibration include bent shafts or gears, damaged bearings, loose bolts, and misaligned couplings or bearings. In compressors, a common cause of excess vibration is an imbalance in the rotating parts.

Periodically checking engines for such problems saves time, money, and energy. Recalibrating devices per manufacturer's recommendations, lubricating on schedule, and checking gauges also help to ensure the most efficient operation. Most manufacturers provide a recommended maintenance schedule. It is best to follow the manufacturer's recommendations for maintaining engines at peak operating efficiency.

7.2.0 Driver Surge

Temporary conditions in fuel or air systems can cause the driver that runs the compressor to surge. It is important to recognize the signs of a surge. Drivers can surge at both low and high speeds. A surge at a low speed may show up in the driver's failure to accelerate. Exhaust temperatures increase and there is a sound of fluttering air. A surge at a high speed results in a loud roar or popping noises. The driver may fail to accelerate to its rated speed. If a surge occurs, shut down the driver immediately to prevent a rapid increase in temperature. After the driver has slowed to a stop, investigate the problem and notify the appropriate personnel.

8.0.0 ◆ TESTING REMOTELY CONTROLLED SHUTDOWN DEVICES

Remotely controlled shutdown devices (RCSDs) are electronic devices that are used to shut down the compressor driver. Each RCSD must be inspected and tested at least once each calendar year to ensure that it is working properly (*49 CFR 192.731*).

Personnel must know how to do the following tasks:

- Ensure that each RCSD is receiving electrical power.
- Test high-temperature shutdown devices.
- Test high-discharge-temperature shutdown devices.
- Test high-liquid-level shutdown devices.
- Test excessive-vibration shutdown devices.
- Test engine-overspeed shutdown devices.
- Test inadequate-lubrication shutdown devices.
- Test high- and low-suction-pressure shutdown devices.
- Properly document the maintenance inspection.

To inspect and test RCSDs, personnel must have a thorough knowledge of the following:

- The company safety procedures
- The company procedures for responding to emergencies
- The shutdown and safety devices being inspected
- The procedures for performing inspections
- The **piping and instrument diagrams (P&IDs)** associated with the equipment being inspected

8.1.0 Testing Procedure

Before testing the RCSD, employees must review the testing procedure. This review is important to ensure the integrity of the system, personal safety, the safety of others, and proper inspection. Always wear the appropriate PPE during the test.

Before the test begins, all affected personnel and the control center must be informed that a station is preparing to test an RCSD. The control center must also be notified when the test is completed. The proper sequence for conducting the test follows:

Step 1 Activate the RCSD.

Step 2 Inspect the RCSD to ensure that all safety, shutdown, alarm, and isolation equipment is working properly, as described on the P&IDs.

Step 3 Make any necessary repairs, or report that repairs are needed.

Step 4 When the RCSD inspection is completed and the RCSD is reset and ready to return to service, notify affected personnel and the control center.

Step 5 Submit the required reports.

8.2.0 Abnormal Operating Conditions

During an RCSD test, there may be a gas leak. Personnel should isolate the source of the leak if it is safe to do so. In addition, personnel must carry out established company safety procedures to protect people, property, and the environment. Once conditions are safe, the leak may be repaired. During repairs, the air must be continuously monitored for hazardous gas.

9.0.0 ◆ MONITORING SAFETY DEVICES

Safety devices are an essential part of monitoring pipeline operations. Some devices are designed to detect changes in pressure or the absence of pressure. Other devices are designed to detect hazardous atmospheres. Following are the types of safety devices found in pipeline operations:

- Hazardous atmosphere devices
- Detection systems
- Gas detection devices
- Smoke, heat, and flame detection devices

When a safety device alarm goes off, it must never be ignored or turned off until the situation has been addressed. Generally, the control center confirms the alarm and notifies field personnel. Field personnel may be directed to investigate alarms to ensure that alarm equipment is working properly.

9.1.0 Hazardous Atmosphere Devices

At some compressor stations, lower explosive limit (LEL), thermal, and **ultraviolet (UV) detectors** continuously sample the air for hazardous atmosphere, high heat, and flames. When the detectors sense an abnormal condition, they activate warning lights and alarms.

9.2.0 Detection Systems

Detection systems (see *Figure 3*) detect fire and unsafe levels of hazardous gas. These systems also protect field equipment and facilities. These systems sound alarms, discharge **fire suppression material,** and shut down equipment. Fire detection and suppression systems protect by doing the following:

- Monitoring hydrocarbon gas levels
- Detecting smoke, flames, and excessive temperatures
- Alerting personnel to a fire or hazardous atmosphere

- Activating the emergency ventilation system
- Isolating hydrocarbon sources
- Automatically shutting down equipment
- Automatically blowing down large-volume natural gas systems
- Automatically discharging fire suppression material

104F03.EPS

Figure 3 ◆ Fixed gas detection system.

In addition, personnel are able to manually discharge fire suppression material by using pull switches. They may do this if the automatic system fails to work.

9.3.0 Gas Detection Devices

Combustible gas detection devices (see *Figure 4*) provide early warning of hazardous atmospheres so that personnel can take appropriate safety steps. By law, the devices must signal an alarm when gas levels reach 25 percent of the LEL. Most alarms are set to warn at 20 percent of the LEL. When gas levels reach 60 percent, equipment shuts down and fire suppression material is dumped. If a monitor is reading close to the LEL and suddenly drops to 0 percent, it may indicate that the area has filled with gas. No alarm can be reset until trained personnel wearing **self-contained breathing apparatus** investigate the condition.

9.4.0 Ionization Detectors

Ionization detectors (see *Figure 5*) sense combustion even before smoke or flames appear. Ionization detectors work well for detecting fire in electrical conduit cable trays. They can detect smoke in facility buildings, crawl spaces, generator rooms, control rooms, and areas containing a lot of electrical equipment.

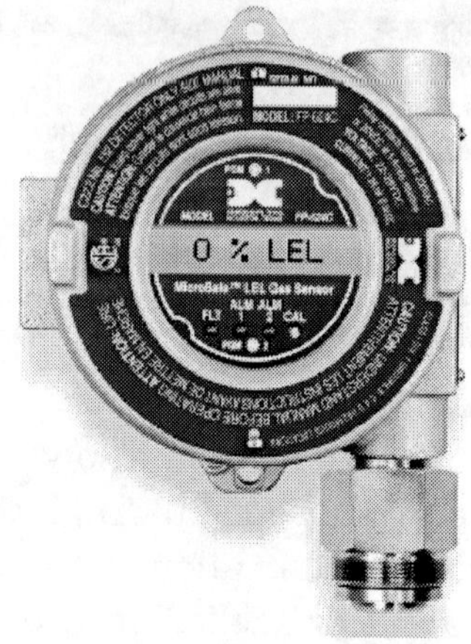

104F04.EPS

Figure 4 ◆ Combustible gas sensor.

104F05.EPS

Figure 5 ◆ Ionization detector.

9.5.0 Ultraviolet Detectors

UV detectors sense fires almost instantly, even when the flames have barely started. In addition, these detectors can distinguish between the UV rays produced by fire and the rays produced by natural light.

That same sensitivity, however, can also cause false alarms. To prevent false alarms, at least two different, overlapping sensors (**voted systems**) must sound an alarm before personnel can act. Causes of false alarms include welding, electrical arc sources, camera flashes, and X-rays or gamma rays.

UV systems work best when nothing gets between the sensor and the fire. Fires that produce heavy smoke make it harder for these systems to detect a fire.

9.6.0 Thermal Detectors

Thermal detectors (see *Figure 6*) react to rapidly rising temperatures by closing controls that are normally open. Thermal detectors are used with graduated temperature settings. This spread of settings allows for special applications. Thermal detectors may be adjusted for a particular setpoint. An early warning sensor closes the controls and sounds an alarm before the setpoint is fully reached.

104F06.EPS

Figure 6 ◆ Thermal detector.

10.0.0 ◆ PIGGING

A **pig** is a device that is inserted into a pipeline to perform a variety of functions that cannot be performed by personnel. Most pigs are maintenance devices, but pigs serve many functions other than cleaning the inside of a pipeline. They also push out water or air during testing, remove liquid from gas pipelines, and monitor a pipeline's interior condition.

The size of the pig depends on the size of the pipeline for which it is designed. Some pigs are as much as 42 inches in diameter. Lengths vary from a few inches to several feet (see *Figure 7*). Pigs are always slightly larger in diameter than the pipeline they are to run in, so that the cups on their surfaces are compressed against the pipeline's walls. This compression aids in the pig's ability to scrape material from the pipeline walls.

Pigs aid in the removal of debris, such as rust, scale, or wax. Buildup of such debris encourages corrosion and restricts the efficiency of the pipeline. Each pig is selected for the specific job that is needed in the pipeline. Using pigs keeps corrosive damage to a minimum and keeps pipeline flow as free as possible. If a pipeline is restricted, sending a pig through the line usually restores full or near-full flow. Pipelines that transport clean products, such as natural gas, usually require fewer pigging operations than pipelines that transport more contaminated products, such as crude oil.

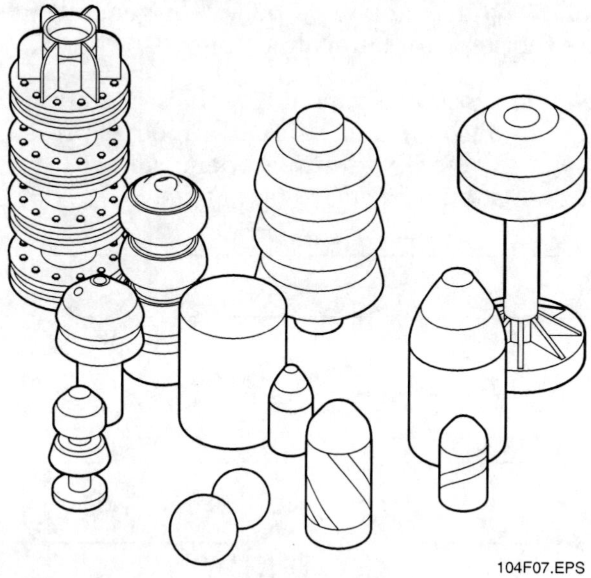

104F07.EPS

Figure 7 ◆ Assortment of pigs.

Smart pigs can perform all kinds of electronic surveillance, including an exact measurement of the pipeline's inner surfaces. They can check the thickness of pipeline walls and find interior damage. Electronic signals transmitted to operations personnel make it easy to track, direct, and retrieve smart pigs.

10.1.0 Launching, Tracking, and Receiving Pigs

Pigging operations are scheduled as regular maintenance on most pipelines. The pigs travel with the product flow and are pushed along by pipeline pressure. They travel much like a piston in an endless hydraulic cylinder. As they travel, they scrape, brush, and clean.

10.1.1 Launching

The methods for **launching** a **pig run** are the same for any type of pig, but they vary with the size of pig. Small pigs are loaded by hand. Pigs for large-diameter pipelines require several employees or, preferably, a hoist or crane. Pig launch procedures vary from one company to another, and you must follow your company's policy. In general, these are the steps for launching a pig:

Step 1 Close all launching valves, isolating the pig launcher from all pipeline pressure. Valve 3 should be open. Valves 1 and 2 should be closed (see *Figure 8*).

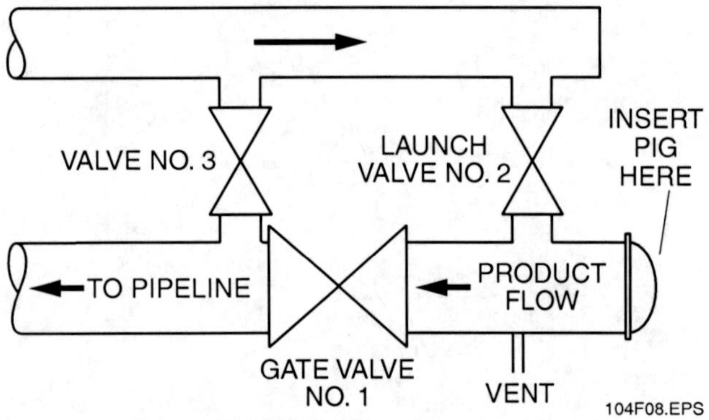

Figure 8 ◆ Typical pig launcher.

Step 2 Open the launching vents. This allows the gas in the section to vent to the atmosphere.

Step 3 Open the launcher end. The end is equipped with a safety pressure-release device that must be opened first in order to release the end cover. The release devices work differently depending on the manufacturer.

Step 4 Load the pig into the barrel.

Step 5 Repressurize the launcher by opening the launching valves until the pressure is equalized. At this point, the pig is ready to be launched.

Step 6 When the signal to launch the pig is received, simultaneously open valves 1 and 2 and close valve 3 to launch the pig.

Step 7 Open valve 3, then close valves 1 and 2 after the pig has moved into the main flow of the pipeline. This step isolates the pig launcher again.

Step 8 Record all procedures on the appropriate report form.

10.1.2 Tracking the Pig

Once the pig has been launched, operations personnel downstream must be told about the type of pig, when it was launched, and roughly when it will arrive. Most stations have some kind of remote device to signal that the pig has arrived.

Pig tracking is vital to the operation of the pipeline because trouble can occur quickly. Pigs can get stuck or hung up in the pipeline, or become so damaged that they cannot travel or can travel only at a reduced speed. Such pigs must be found immediately to prevent serious problems in the pipeline.

If a problem develops, operations personnel may notice a sudden change in pressure. The pig may be late or fail to arrive. When personnel are certain that the pig has not arrived, they must notify the control center immediately. Some pigs are equipped with electronic signaling devices that help pinpoint their location. For pigs without signaling devices, a search must be initiated.

A typical search begins by checking all station piping and equipment. If the pig is hung in a valve, personnel may hear a noise from inside the pipeline or valve. If the pig has been damaged or has self-destructed, pieces of the pig may be found in the strainer.

If the pig is not in the station piping, the search must move upstream. The first place to look is the launcher at the original station to ensure that the pig left the station. If the pig is not found upstream, the search must move to the downstream section of the pipeline.

Once the pig is located, a second pig may be launched in an effort to bump or push the first pig loose. The danger is that the second pig may also become stuck. The second pig should be radio equipped so that its progress may be followed closely. If the second pig is able to push the first one loose, the two must be monitored closely to ensure that both pass through the pipeline.

10.1.3 Receiving and Unloading the Pig

Once the pig has been launched, information about its type and schedule must be sent to operations personnel downstream. A detector signal should be sent to the control center of the next station, warning that the pig is approaching. The valve alignment must be prepared to receive the pig (see *Figure 9*).

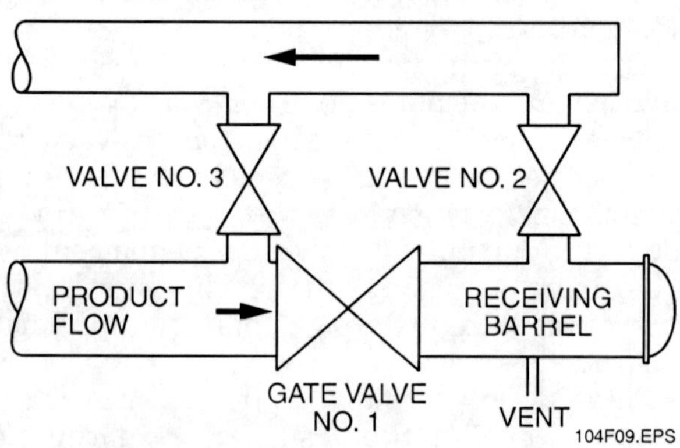

Figure 9 ◆ Pig-receiving installation.

The procedure for receiving and unloading pigs may vary from one company to another, and you must always follow your company's procedures. In general, these are the steps that must be followed when receiving a pig:

Step 1 Set up the valve alignment to receive the pig by opening valves 1 and 2 and closing valve 3.

Step 2 Listen for the pig to arrive at the receiving unit.

Step 3 Ensure that the restraining bar is in place to keep the pig from going back into the pipeline.

Step 4 Open valve 3, and close valves 1 and 2.

Step 5 Vent gas to the atmosphere.

Step 6 Release all pressure before attempting to open the receiver end cap.

Step 7 Pull the pig from the receiving barrel.

Step 8 Place the pig on a concrete slab for inspection.

Step 9 Inspect the pig.

Step 10 Complete required reports and send them to the control center. The reports should document the pig's condition and all test results.

Review Questions

Sections 6.0.0–10.0.0

1. To keep gas compressed and moving through the pipeline, compressor stations are necessary every _____ miles or so along the pipeline.

a. 50
b. 100
c. 200
d. 500

2. Two common problems that occur on engines are _____.

a. excess vibration and driver surge
b. missing starters and bent springs
c. jams and freezes
d. broken bolts and sheared pins

3. A(n) _____ is an electronic device that can be used to shut down the compressor driver.

a. smart pig
b. RCSD
c. manifold
d. pressure-relief valve

4. _____ react to rapidly rising temperatures by closing controls that are normally open.

a. UV detectors
b. Flame detectors
c. Thermal detectors
d. Smoke detectors

5. Besides cleaning the inside of a pipeline, pigs can also _____.

a. perform minor repairs
b. monitor the interior condition of the pipeline
c. odorize the gas in the pipeline
d. remove odorant from the pipeline

11.0.0 ◆ MONITORING WEATHER CONDITIONS

Many pipeline operations depend on moderate weather. A tornado, hurricane, or blizzard can threaten personnel and the pipeline itself. High winds, snow, ice, and cold temperatures may also hinder operations. Sometimes the approach of an electrical storm signals the danger of deadly lightning and wind.

The weather must be monitored to protect the safety of employees working on the pipeline. In addition, company procedures should be followed for properly recording all weather changes.

Trying to continue normal operations outside during inclement weather could result in serious injury or death. Personnel may alter activities and seek shelter as warranted. In severe cases, the pipeline may have to be shut in. When control of the pipeline can be ensured and it is safe for personnel to return to work, operations can resume. Company procedures should be followed when determining the correct response to weather conditions.

The sooner that personnel become aware of a breaking weather situation, the more likely they can take action before problems become critical. To monitor weather changes effectively, personnel sometimes rely on weather-reporting Web sites. Many stations use small weather monitors to gauge local conditions. Following are examples of some conditions that are monitored:

- Temperature
- Wind direction and speed (with high-wind alarm)
- Wind chill factor (with low-wind-chill alarm)
- Daily rainfall
- Barometric pressure

12.0.0 ◆ LOCATING PIPELINES

It is important to accurately locate pipelines before doing any excavation. Digging in the wrong place wastes time and energy and may cause damage to the pipeline. The pipeline's location should be clearly shown on company alignment maps, which should be used with pipeline markers to locate the pipeline.

However, your company may not be the only pipeline or utility with buried lines in the area, so the One-Call notification system must be alerted before any excavation is planned. When the system is notified, companies with buried lines in the area are alerted and asked to provide temporary markings, such as flags, stakes, or paint. These companies should provide line depths and descriptions as well.

Once the correct pipeline, called the target pipeline, is located, a backhoe may be used to quickly remove large amounts of ground cover. The backhoe operator must take care when operating this powerful equipment. Digging too deeply or off center can damage the pipeline. Backhoe excavating is performed in stages. The first stage often destroys the temporary markings that show the position of the pipeline. Therefore, during excavation, the line must be periodically relocated and re-marked to help the backhoe operator avoid damaging the line.

12.1.0 Alignment Maps

Alignment maps are used to help locate the target pipeline or other buried lines. Alignment maps can be of several different types and show the location, direction, and depth of underground lines. Individuals responsible for locating the target line and other buried lines must be skilled in reading alignment maps. If there is a question about the accuracy of the maps, a surveyor's map or blueprint should be obtained. In cases where no other information is available, it may be necessary to obtain the as-built specifications. As-built specifications are drawings showing installations as they were built.

12.2.0 Mechanical Locators

Personnel may locate the target pipeline and other buried lines mechanically by probing, **pot-holing** by hand, and using vacuum units. The method chosen depends on the circumstances; in some cases, methods may be combined.

12.2.1 Probing

Personnel use a **probe** (see *Figure 10*), which is a steel or fiberglass rod, to determine the depth of buried lines and cables. The rod has a smooth, rounded end and generally has a T-shaped handle. To use a probe, personnel slowly push it into the ground until contact is made with the pipeline. Probe rods must be used carefully to avoid breaking the pipeline coating and should be used only by trained and qualified personnel.

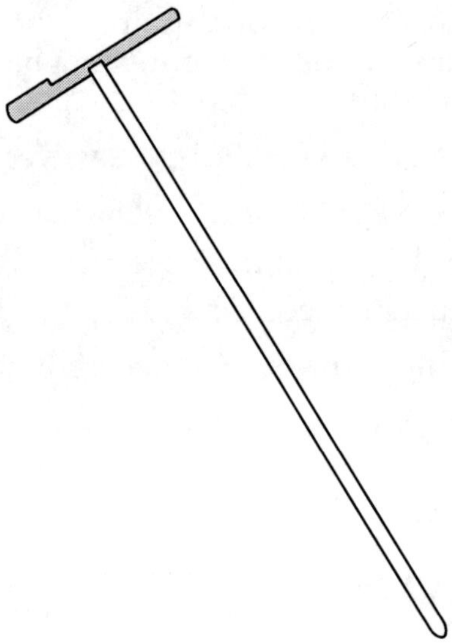

104F10.EPS

Figure 10 ◆ Typical probe.

WARNING!
Proceed slowly and carefully when probing. An electrocution hazard from electrical lines may be present. To offset this risk, use only tools made of a nonconductive material, such as fiberglass, or tools equipped with insulated handles.

NOTE
Some companies do not allow probing at all. Check your company's policies.

12.2.2 Hand Digging and Pot-Holing

Pot-holing is a method used by personnel to locate the pipeline by hand-digging small holes. Pot-holing may be more hazardous than probing because the sharp edges of the tools used could damage the pipeline. Personnel using this method must dig slowly and carefully and work with tools that have insulated handles.

WARNING!
Proceed slowly and carefully when pot-holing. Sharp-edged digging tools can cut electrical lines, creating an electrocution hazard. As when probing, use only tools made of a nonconductive material, such as fiberglass, or tools equipped with insulated handles.

12.2.3 Using Vacuum-Powered Tools

Vacuum-powered tools are designed to reduce the risks associated with exposing a buried pipeline. They come in two basic types: water-vac (see *Figure 11*) and air-vac.

104F11.EPS

Figure 11 ◆ Water-vac truck.

Water-vac units use water to hydraulically bore a hole. Once the hole is bored, a vacuum on the unit is used to remove the ground cover. A water-vac opens a smooth, clean hole with no damage to the line.

Air-vac units operate on the same principle as water-vac equipment. Air-vac units also vacuum out the ground cover, but bore holes using compressed air instead of water.

WARNING!
Vacuum-powered tools propel dirt and rock particles into the air. You must wear eye protection when using vacuum-powered tools.

12.3.0 Electronic Locators

Electronic locators are lightweight and portable. These locators are accurate, but company policy may require that a mechanical location tool also be used to confirm the location before excavation can take place. The procedures for using electronic locators vary depending on the type of locator used. For exact information, it is best to read and follow the manufacturer's recommendations. In addition, your company may have specific procedures for this process. Generally, field personnel follow these steps to locate a buried pipeline using an electronic locator:

Step 1 Review maps, alignment sheets, and as-built specifications.

Step 2 Look for signs of other buried lines such as valves, valve box covers, manhole covers, pedestals, junction boxes, or utility poles with downward cables for clues of what might be underground.

Step 3 Sweep the area for indications of buried lines.

Step 4 Locate and mark the location of buried lines.

Step 5 Verify the location of the target pipeline.

Step 6 Determine the depth of the target pipeline.

Step 7 Repeat the process if required to confirm your findings.

Step 8 Report your findings to the appropriate personnel.

12.3.1 Stick Locators

Stick locators (see *Figure 12*) are also known as metal detectors or ferromagnetic locators. They have two sensors encased in a shaft, which detect a magnetic field near underground items that contain iron, such as pipelines, survey pins, and tanks. Because these locators detect only objects containing iron, they generally are not sensitive to electrical wiring, telephone lines, or cables. Some models have a feature that indicates the presence of live, flowing electric current. In addition, some models include an erase feature that allows the operator to screen out small iron-containing items, such as nuts and bolts, from the readings.

The procedures for using stick locators vary depending on the type of locator used. For exact information, it is best to read and follow the manufacturer's recommendations. In addition, your company may have specific procedures for this process. Generally, field personnel follow these steps to locate a buried pipeline using a stick locator:

Step 1 Walk steadily across the target area, sweeping the locator slowly back and forth above the ground.

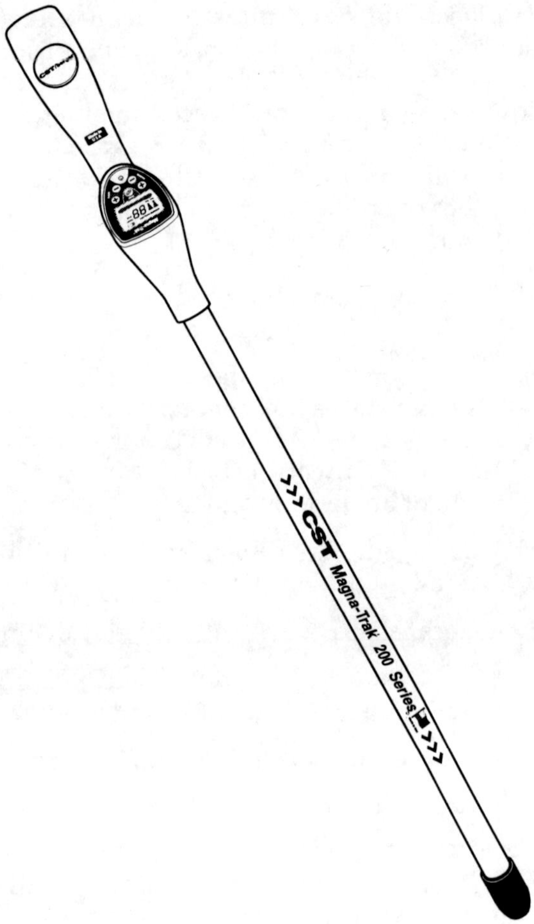

104F12.EPS

Figure 12 ◆ Typical stick locator.

Step 2 Move with the locator in a wide circle (about 6 feet out) around the target area.

Step 3 Hold the locator upright while walking in a smaller circle to narrow the search and pinpoint the location of the buried item.

Step 4 Repeat the process if required to confirm your findings.

Step 5 Report your findings to the appropriate personnel.

> **WARNING!**
> Stick locators may locate both live and dead lines. Always assume a line is live until you confirm otherwise.

> **CAUTION**
> Do not use a stick locator while wearing steel-toed boots. The iron content in the steel may throw off readings.

12.3.2 Electromagnet (Battery-Operated) Locators

Electromagnetic locators range from simple, portable units that can detect buried pipelines and cables to complex units that can identify buried systems, system faults, and the depth of system components. These locators contain two main components: a transmitter and a receiver.

The transmitter sends an easily identifiable signal to locate the buried pipeline. This signal is transmitted to the receiver, which provides a reading showing the depth of the buried line. Transmitter signals operate over a wide range of frequencies. The signal power ranges from a low of between 0.1 and 5 watts to a high of 150 watts. The more powerful the signal, the greater the distance over which it can be detected.

The procedures for using electronic locators vary depending on the type of locator used. For exact information, it is best to read and follow the manufacturer's recommendations. In addition, your company may have specific procedures for this process. Generally, field personnel follow these steps to locate a buried line using an electromagnetic locator:

Step 1 Select the frequency recommended in the operating instructions.

Step 2 Make adjustments and choose settings as recommended by the manufacturer.

Step 3 Apply the transmitter signal to the target area.

Step 4 Sweep the area, keeping the locator in a vertical position.

Step 5 Mark the point of the strongest signal.

Step 6 At the point of the strongest signal, rotate the locator 90 degrees until the minimum signal is achieved to locate the target pipeline.

> **NOTE**
> Electronic locators transmit signals from either side of the line. The minimum signal indicates that the operator is directly over the target pipeline.

Step 7 Place the receiver on the ground and press the depth key to display the target depth.

Step 8 Repeat the process if required to confirm your findings.

Step 9 Report your findings to the appropriate personnel.

> **NOTE**
> All locators sense depth to the center of the signal, roughly the center of the pipeline. To obtain the cover depth, subtract ½ of the pipeline diameter from the depth measured between the spots to get actual cover depth.

13.0.0 ◆ UPRATING THE PIPELINE MAOP

Uprating is the term used to describe the process for increasing the MAOP on an existing pipeline segment. The MAOP is the highest pressure allowed for a pipeline segment. Various factors, such as pipeline design and operating experience, are considered when setting the MAOP. Engineers who have been trained on pipeline design and safe operating limits handle uprating the MAOP. However, it is useful for all pipeline personnel to have a basic understanding of this process.

Step-by-step pressure increases are necessary to safely uprate a pipeline. A trained individual making such pressure increases must know a great deal about pipeline operations. This knowledge includes the following areas:

- A thorough knowledge of the uprating plan
- A basic knowledge of the system being uprated
- An understanding of company safety policies and procedures

13.1.0 Performing Step-by-Step Pressure Increases

Before performing step-by-step pressure increases, it is essential to increase the pressure in the pipeline segment to the current system MAOP. Doing so provides a starting point for the step-by-step pressure increases. Then the pipeline pressure must be increased gradually to the first planned setting.

When the pipeline pressure reaches the first planned setting, the process must halt before the pressure is increased to the next setting. Gradual, step-by-step pressure increases help to prevent pipeline failure and accidents. Each pressure increase must be recorded on a pressure test form.

The pipeline must be monitored constantly as the pressure increases. A pressure loss could indicate a leak. If there is a pressure loss, a leak survey must be conducted to check for possible leaks. Leaks must be monitored or repaired as specified in *49 CFR Subpart K*.

Step-by-step increases in pipeline pressure should be continued, with a stop at each setting as planned, until the final planned test pressure is reached. When the final pressure is reached, the pipeline is uprated.

13.2.0 Documenting the MAOP

The final established MAOP must be documented. The following information must be included:

- The name of the operating company
- The name of the contractor
- The name of the person conducting the uprating test
- The date and time of the test
- The location of the pipeline or the pipeline segment uprated
- The test medium used
- The test pressures
- The duration of the test
- The uprating plan
- Any leakage and repairs
- Any AOCs and the corrective actions taken

13.3.0 Abnormal Operating Conditions

If a leak occurs during an uprating, the test must be stopped and a determination made whether to repair the leak or continue the test while monitoring the leak. Each detected leak must be repaired or monitored in accordance with DOT regulations (*49 CFR Subpart K*). Personnel must carry out company safety procedures to ensure the safety of those who are performing the uprating procedure.

14.0.0 ◆ OPERATING ODORIZATION EQUIPMENT

Natural gas is odorless. For safety reasons, a federal regulation requires that a person with a normal sense of smell must be able to easily detect natural gas when the gas concentration is only 20 percent of its LEL (*49 CFR 192.625*). Therefore, an odorant, usually a chemical called mercaptan, is added to the natural gas stream in distribution lines. Generally, local distribution companies are responsible for adding the odorant, which gives the gas a distinctive rotten-egg smell.

Personnel responsible for injecting odorant into a natural-gas stream should read and follow the material safety data sheet (MSDS) for the odorant being used. In addition, employees must follow established company procedures for injecting odorant and wearing the appropriate PPE.

The steps for injecting odorant may vary from one company to another. Generally, these are the basic steps for operating odorization equipment:

Step 1 Ensure that the odorant is the proper odorant for the intended use.

Step 2 Operate the odorization equipment so that the odorant is injected smoothly and consistently.

Step 3 Measure the odorant level in the pipeline system.

Step 4 Monitor the odorant level in the pipeline system.

Step 5 Submit the required reports.

14.1.0 Injecting the Odorant

The amount of odorant to be injected can vary from place to place, even within the same system, depending on operating conditions. Following are three types of equipment that may be used to inject the odorant:

- *Bypass type*—Diverts the main gas stream through a mechanical device to disperse the liquid odorant into the gas
- *Wick type*—Uses the flow of gas across a wick or **pitot tube** to pick up liquid odorant for dispersion in the gas stream
- *Electronic*—Uses an onboard microprocessor to provide a flow measurement signal and generate data that is stored in an audit trail

14.2.0 Monitoring the Odorant

Once the odorant has been injected into the gas, personnel must monitor it to ensure that proper ratios are maintained. The type of equipment used for this purpose depends on company policy. Generally, personnel use an **odorometer** (see *Figure 13*). This meter allows personnel to determine if the ratio of the usage to the volume of gas is within acceptable limits. Odorometers must be properly calibrated to produce valid measurements. Personnel must follow the manufacturer's recommendations for proper use.

15.0.0 ◆ PERFORMING ROW INSPECTIONS

Because pipelines cover so many miles, companies must use several methods to survey the pipeline (*49 CFR 192.705*). These methods include walking, driving, and flying. In addition, remote video equipment allows personnel to monitor certain areas. Gas pipelines must be patrolled regularly (see *Table 1*).

In addition to regular patrols, maintenance crews performing routine tasks in the field must also be alert to potential problems. Experienced workers recognize the normal sounds and smells of their workplace. Maintenance personnel should recognize safety-related conditions and report them to the appropriate supervisor.

Inspection personnel check the pipeline ROW for any conditions affecting the integrity of the pipeline. Some personnel use computers to record information and send it to the control center electronically. This equipment may be used to produce summary reports and pinpoint errors (see *Figure 14*).

Personnel check for a variety of conditions along the pipeline ROW. Following are some of the conditions they look for:

- Construction work
- Soil erosion or growth of vegetation
- Missing or damaged pipeline markers
- Encroachment
- Signs of vandalism

15.1.0 Monitoring by Aircraft

Aerial patrols, which are the most common form of monitoring, are performed on a schedule set by the pipeline company. Using current maps of the pipeline, pilots fly slowly at a safe altitude to allow for a good view. They radio their reports to the control center. Most companies require aerial patrols to keep a written record of their observations. In some cases field personnel are sent to investigate an aerial report and complete a follow-up report. Pilots may be required to review any follow-up reports and compare them to the original patrol log.

15.2.0 Monitoring by Vehicle and Foot Patrol

Although aerial patrols are more commonly used to inspect the pipeline, experienced ground patrols, called **linewalkers,** may be required to inspect areas where low-level flying is prohibited. In addition, linewalkers may be required when visibility is low or when bad weather grounds aircraft. In some cases, linewalkers are sent to investigate an aerial report more closely. Linewalkers may inspect the line on foot or may drive along areas that are accessible by car.

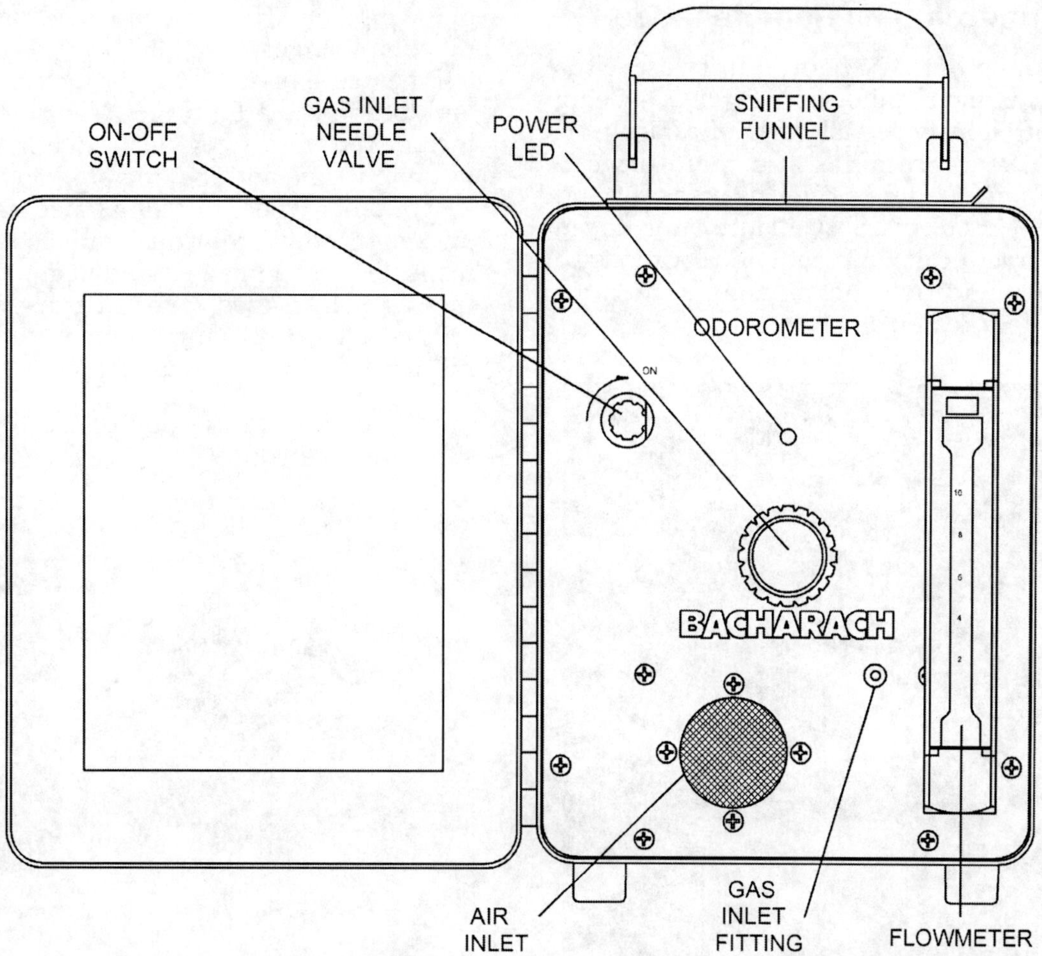

104F13.TIF

Figure 13 ◆ Odorometer.

Table 1 Maximum Interval Between Patrols

Class Location of Line	Maximum Interval between Patrols	
	At Highway and Railroad Crossings	**At All Other Places**
1 and 2	7½ months, but at least twice each calendar year	15 months, but at least once each calendar year
3	4½ months, but at least four times each calendar year	7½ months, but at least twice each calendar year
4	4½ months, but at least four times each calendar year	4½ months, but at least four times each calendar year

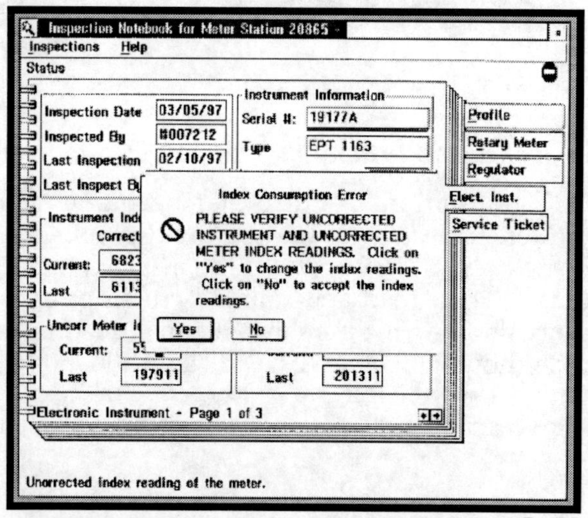

104F14.TIF

Figure 14 ◆ Sample electronic instrument inspection page with error message.

15.3.0 Monitoring by Remote Video

Some facilities provide 24-hour security, as well as video cameras and monitors (see *Figure 15*). Being able to continuously see all parts of a facility is important. Video cameras and monitors can reveal equipment failure as well as break-ins and other security problems. Surveillance cameras can also focus on critical equipment and locations so that security and field personnel can monitor these areas for problems.

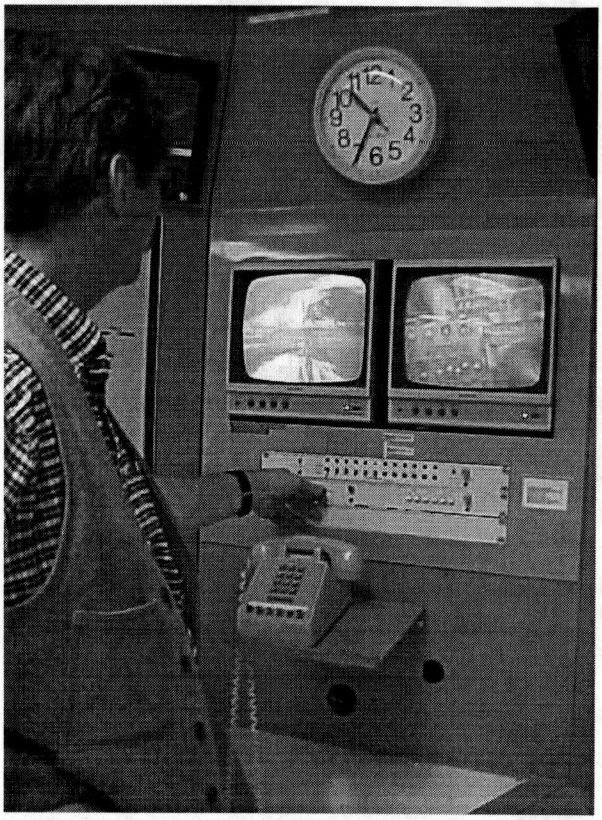

104F15.EPS

Figure 15 ◆ Typical video monitors.

15.4.0 Line Markers

DOT regulations require that permanent line markers show where underground pipelines are located (*49 CFR 192.707*). These markers must be located at all road and railroad crossings and at intervals along the pipeline. The markers must be close enough together to clearly show the position of the pipeline. They also must indicate where the pipeline changes direction.

Line markers must display the company name and location (see *Figure 16*). The signs must also include the word *WARNING*, identify the product in the pipeline and the operator, and give a toll-free number where the company can be called in an emergency. Markers must be routinely inspected for damage, vandalism, and illegible messages. Damaged or missing markers must be replaced. The following is one procedure used to inspect and maintain line markers:

Step 1 Identify the section of the pipeline to be inspected.

104F16.EPS

Figure 16 ◆ Line marker.

Step 2 Locate the most recent maps (see *Figure 17*), records, and alignment drawings, and use them to accurately locate the section of pipeline to be inspected.

Step 3 Determine the access requirements for the section to be inspected. Some ROWs are restricted.

Step 4 Inspect the line markers and replace any that are missing or damaged or contain incorrect information. In addition, ensure that there are enough signs and that they have been put in the right places.

Step 5 Submit the required reports.

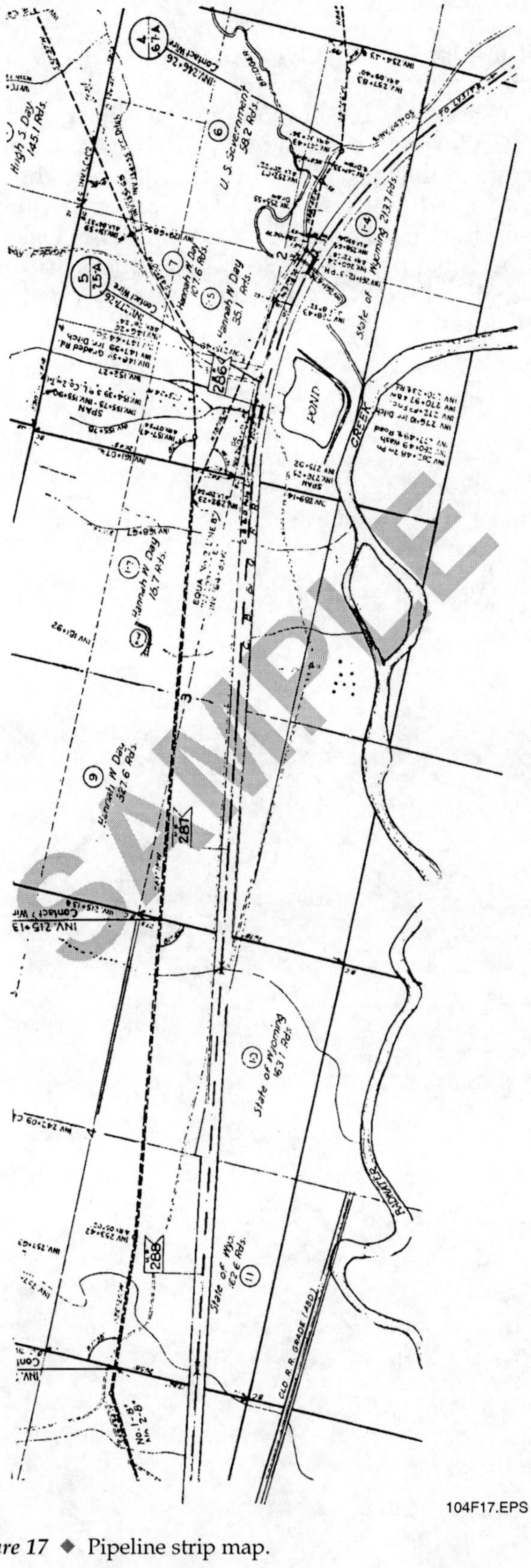

104F17.EPS

Figure 17 ◆ Pipeline strip map.

Review Questions

Sections 11.0.0–15.0.0

1. _____ are also known as metal detectors.

a. Electromagnetic locators
b. Stick locators
c. Probe locators
d. Sensor locators

2. When _____ a pipeline, an essential step is to increase the pressure to the current system MAOP.

a. purging
b. uprating
c. pigging
d. cleaning

3. Natural gas is usually odorized by _____.

a. the local gas distribution company
b. the pipeline transmission company
c. control center personnel
d. field personnel

4. Personnel who inspect a pipeline's right-of-way on the ground, instead of by air, are called _____.

a. batch trackers
b. ROW walkers
c. linewalkers
d. line trackers

5. Line markers must display all of the following *except* _____.

a. the word *WARNING*
b. the product in the pipeline
c. the name of the pipeline operator
d. the distance to the next line marker

Summary

In this module, you learned that routine field operations are often anything but routine. Working in the field requires you to learn how pipeline equipment works and how to deal with problems that arise. You must know what conditions are normal so that you can quickly spot any condition that is not normal. Fieldwork can be challenging and sometimes dangerous. Although natural gas is generally regarded as safe, it can be a hazardous substance under certain conditions. Transporting it in pipelines involves some risks for pipeliners, the public, and the environment, which is why you must always do your job safely and responsibly.

This module provided you with basic knowledge of the tasks you must perform and the skills you need for routine field operations. Although you may not have to perform all of the tasks outlined in this module, a basic understanding of what is involved will help you to perform your own tasks better and more safely.

Trade Terms Introduced in This Module

Abnormal operating condition (AOC): A condition within a pipeline system that exceeds operating design limits or that may result in a hazard to persons, property, or the environment (*49 CFR 192.803*).

Blowdown: The depressurization of gas piping, which is accomplished by releasing the pressure at a safe atmospheric vent location.

Bonding: The permanent joining of metallic parts to form an electrically conductive path, which ensures a safe electrical connection.

Combustible gas indicator (CGI): An instrument used to sample the atmosphere to determine if it is explosive.

Compressor station: Facilities installed along pipelines approximately every 100 miles that compress natural gas to reduce its volume and force it through the pipeline.

Driver: A mechanical component that supplies power to move another mechanical component.

Fire suppression material: A chemical or gas used to smother fires quickly.

Hot tap: A process in which a drill or circular cutter is used to cut a hole in a pipeline in which gas is still flowing to shut in, or isolate, that pipeline segment.

Hot work: Any activity that could produce flames, sparks, or other hot fragments that might act as an ignition source to flammable materials in the area. The activities covered by this term include, but are not limited to, welding, flame cutting, and soldering.

Ionization detector: A photoelectric detection device that signals an alarm when it detects smoke.

Launching: Releasing a pig into a pipeline. It is also known as kicking.

Linewalker: A member of the ground patrol that inspects a pipeline's right-of-way.

Lower explosive limit (LEL): The minimum concentration of vapor in air below which fire does not occur in the presence of an ignition source.

Manifold: A pipeline fitting with several side outlets to connect it with other piping. It is also known as a header.

Odorometer: An instrument that is used to measure the odorant level in a gas.

Pack: To compress natural gas under pressure in a pipeline.

Pig: A scraping tool with blades or brushes that is forced through a pipeline to clean out accumulations of wax, scale, and debris from the pipeline walls. It is also called a line scraper or go-devil.

Pig run: A pig's trip through a pipeline.

Pig tracking: The process of monitoring the progress of a *pig* as it travels through the pipeline.

Pipeline segment: (1) A specified length of pipeline constructed using a specified grade of piping. (2) A section of pipeline that must be shut in, or isolated, for maintenance or repairs.

Piping and instrument diagrams (P&IDs): Engineered diagrams of piping, valves, and electrical utilities.

Pitot tube: A tube placed in a gas or fluid flow to measure velocity.

Pot-holing: A procedure used to locate underground installations by hand-digging a small hole in the ground.

Probe: A T-shaped hand tool, 4 to 6 feet long, used to locate underground installations by pushing it into the soil repeatedly at regular intervals.

Purge: To replace the atmosphere in a pipeline by injecting an inert substance to prevent formation of explosive mixtures.

Remotely controlled shutdown device (RCSD): An electronic device that can be used to shut down a compressor driver from a control center or other location.

Self-contained breathing apparatus: A device consisting of a face piece and an air canister mounted in a shoulder harness that enables workers to enter areas where there is not enough oxygen or where there are dangerous concentrations of fumes or gases.

Smart pig: An electronic device inserted into a pipeline that senses and records irregularities in the line. It is sometimes called an instrumented pig.

Steady state: The point at which pressures and flows are relatively constant over time.

Thermal detector: A heat-sensing device that warns of the presence of flame by detecting a sudden rise in temperature. It is also known as a flame detector or a bolometer.

Ultraviolet (UV) detector: A sensor that notes the presence of fire by detecting the ultraviolet rays given off by fires that are just beginning.

Unpack: To decompress gas that is under pressure in a pipeline.

Upper explosive limit (UEL): The maximum concentration of vapors in the air that will burn when an ignition source is present.

Uprating: Increasing the maximum allowable operating pressure on an existing pipeline segment.

Valve alignment: The exact sequence of valves that must be opened and closed at the same time so that gas can be received at and delivered to specific locations.

Voted system: An alarm system layout used for flame detection and gas detection, employing sensitive detectors and delayed action relays. Such systems will not alarm until two or more detectors are tripped.

Figure Credits

Gardner Denver	104F01
Matheson Tri-Gas, Inc.	104F03
Bacharach, Inc.	104F13
Pipeline	104F14
Kinder Morgan	104F16

CONTREN™ LEARNING SERIES — USER UPDATES

The NCCER makes every effort to keep these textbooks up-to-date and free of technical errors. We appreciate your help in this process. If you have an idea for improving this textbook, or if you find an error, a typographical mistake, or an inaccuracy in NCCER's Contren™ textbooks, please write us, using this form or a photocopy. Be sure to include the exact module number, page number, a detailed description, and the correction, if applicable. Your input will be brought to the attention of the Technical Review Committee. Thank you for your assistance.

Instructors – If you found that additional materials were necessary in order to teach this module effectively, please let us know so that we may include them in the Equipment/Materials list in the Instructor's Guide.

Write: Curriculum Revision and Development Department
National Center for Construction Education and Research
P.O. Box 141104, Gainesville, FL 32614-1104

Fax: 352-334-0932

E-mail: curriculum@nccer.org

Craft ____________ Module Name ____________

Copyright Date ____________ Module Number ____________ Page Number(s) ____________

Description

____

____

____

____

(Optional) Correction

____

____

____

(Optional) Your Name and Address

____

____

____

Module 67105-02

Routine Control Center Operations

(CT 43, 50, 51, 54, 56, 57, and 58)

Introduction to the Pipeline Operator Qualification Rule, API, and NCCER

It could be said that petroleum products keep America running. Petroleum products like oil and natural gas are the principal sources of energy in the United States and are used in many manufacturing and chemical processes. Petroleum products are primarily transported by pipeline. There are more than a million miles of liquid and natural gas pipeline in this country, ranging in diameter from two inches to four feet. These pipelines bring crude oil and natural gas from production fields to processing plants and refineries—and ultimately to the consumer.

Pipeline Operator Qualification

Petroleum products are considered hazardous, so pipeline personnel must take extra precautions and follow specific procedures when working with them. To help ensure the safe handling and transportation of petroleum products, the United States Department of Transportation (DOT) mandates safe practices as federal regulations.

One such regulation, the Operator Qualification (OQ) rule, requires pipeline operators to develop a written qualification plan to evaluate their personnel and their contract personnel (see *49 CFR Part 192 Subpart N—Gas,* and *49 CFR Part 195 Subpart G—Liquid*). The evaluation hinges on the operator's ability to perform certain covered tasks and to be able to recognize and respond to abnormal operating conditions that may be encountered while performing these covered tasks.

Covered Tasks

A pipeline performance task is deemed *covered* if the activity passes the following four-part test:

1. Is it performed on a pipeline facility?
2. Is it an operations or maintenance task?
3. Is it performed as a requirement of *49 CFR Part 195 (Liquid)* or *49 CFR Part 192 (Gas)*?
4. Does it affect the operation or integrity of the pipeline facility?

If the answer to all four questions is *yes*, then the activity is considered *covered* and must figure into the evaluation process of pipeline personnel.

American Petroleum Institute

The OQ rule and its related covered tasks provided the catalyst to develop a unified pipeline training and assessment program—so that pipeline operators and owners have ready access to the assessment and training tools they need. This effort was the direct result of the vision and leadership provided by The American Petroleum Institute (API) and API members.

Efforts to unify the industry are nothing new to API. Indeed, the impetus for forming the Institute in 1919 was the need to standardize engineering specifications for drilling and production equipment. Today, the most pressing issues revolve around public perceptions and government policies toward the industry. Speaking with one voice on these issues has become as essential as having interchangeable parts in the field. API's members determine the petroleum industry's positions on public policy as well as the standards that govern its day-to-day operations worldwide.

Consortium on Operator Qualification

API assembled a focus group whose primary purpose was to identify covered tasks and provide compliance recommendations to pipeline operators.

The Consortium on Operator Qualification (COOQ) became the guiding force for the identification and development of covered tasks for the liquid pipeline industry. Their work product, API publication 1161, *Guidance Document for the Qualification of Liquid Pipeline Personnel,* formed the basis for the resulting Pipeliner Training and Assessment Program.

The covered task list created in API publication 1161 has now evolved to include gas covered tasks and is maintained by NCCER. The most current covered task list can be downloaded at www.nccer.org.

National Center for Construction Education and Research

With the guidance and support of API, its members, and the Consortium on Operator Qualification, the National Center for Construction Education and Research (NCCER) was able to design and implement the Pipeliner Training and Assessment Program.

NCCER is a not-for-profit 501(c)(3) education foundation established in 1995 by the world's largest and most progressive construction and maintenance contractors and national contractor associations. NCCER was founded to address the

severe workforce shortage facing the construction and maintenance industries and to develop a standardized training process and curricula. Today, NCCER is supported by hundreds of leading construction and maintenance companies, manufacturers, and national associations and developed more than 35 construction and maintenance titles.

NCCER also maintains a National Registry, which contains the records of trainees and instructors who have received NCCER training. All completion information for every trainee is sent to the NCCER and filed within the Registry. The Registry can then confirm training and skills for workers as they move from company to company, state to state, or even within their own company.

Standardized Curriculum (NCCER's Contren™ Learning Series)

NCCER curricula are developed using the expertise of industry subject matter experts (SMEs). The use of SMEs as an integral part of every new development or revision ensures that the curriculum is current, industry-driven, and indicative of industry practices. The development and/or revision process is facilitated from start to finish by NCCER project managers, who work closely with the SMEs and a select team of technical writer(s) on the development and presentation of the curriculum.

Pipeline: Five Training Paths

The pipeline curriculum focuses on essential skills and competencies within five skill areas:

- Corrosion Control
- Electrical and Instrumentation
- Maintenance
- Mechanical
- Operations (Field and Control Center)

SMEs from the pipeline industry have determined the skills and competencies needed based on API's covered tasks and the consensus of the group/industry. Specific company information is generally not incorporated into the curriculum unless the SMEs determine that the information is the industry standard.

Training Modules and Covered Tasks

The smallest unit of instruction in the curriculum is a *module*. Each curriculum is composed of modules (the equivalent of stand-alone chapters) that are logically sequenced into complete *levels* of instruction. Every module is task-oriented and requires the successful completion of a written (knowledge-based) examination and a performance (hands-on) test. Modules may be individually selected and/or combined for efficient task training. Modules in the pipeline curricula, to the extent possible, will align with API-developed covered tasks. While not all modules will relate to covered tasks, all covered tasks will relate to at least one module.

Benefits of Training to the Pipeline Marketplace

Applying NCCER's standardized training process allows pipeline owners and contractors in the United States to establish the benchmark for acceptable skill and competency levels within the pipeline industry. This process will assist training organizations in developing and/or improving standardized pipeline training programs that effectively match their resources and operating conditions while responding to regulatory demands and to the needs of the industry. The process will also recognize training organizations that meet the minimum standards of competence, warranting regulatory, public, and professional confidence. This in turn will foster national unity within the pipeline industry for the mutual benefit of individual technicians, owners, contractors, and ultimately pipeline-users and society.

Additional Resources

To review the Pipeliner Training and Assessment Program Implementation and Resource Guide and the latest version of pipeline covered tasks, visit NCCER's Web site at www.nccer.org, or contact:

National Center for Construction Education and Research
PO Box 141104
Gainesville, FL 32614-1104
(888) 622-3720
Web site: www.nccer.org

For more information on API, please contact:

American Petroleum Institute
1220 L Street, Northwest
Washington, D.C. 20005
(202) 682-8000
Web site: www.api.org

To order *API Pub 1161*, contact Global Engineering Documents by calling 1-800-854-7149 or visiting their Web site at www.global.his.com.

Additional sources of information on the pipeline operator qualification regulation include:

DOT – Office of Pipeline Safety
www.ops.dot.gov

Transportation Safety Institute
www.tsi.dot.gov

COURSE MAP

This course map shows all of the modules in the Gas Pipeline Operations curriculum. The suggested training order begins at the bottom and proceeds up. Skill levels increase as you advance on the course map. The local Training Program Sponsor may adjust the training order.

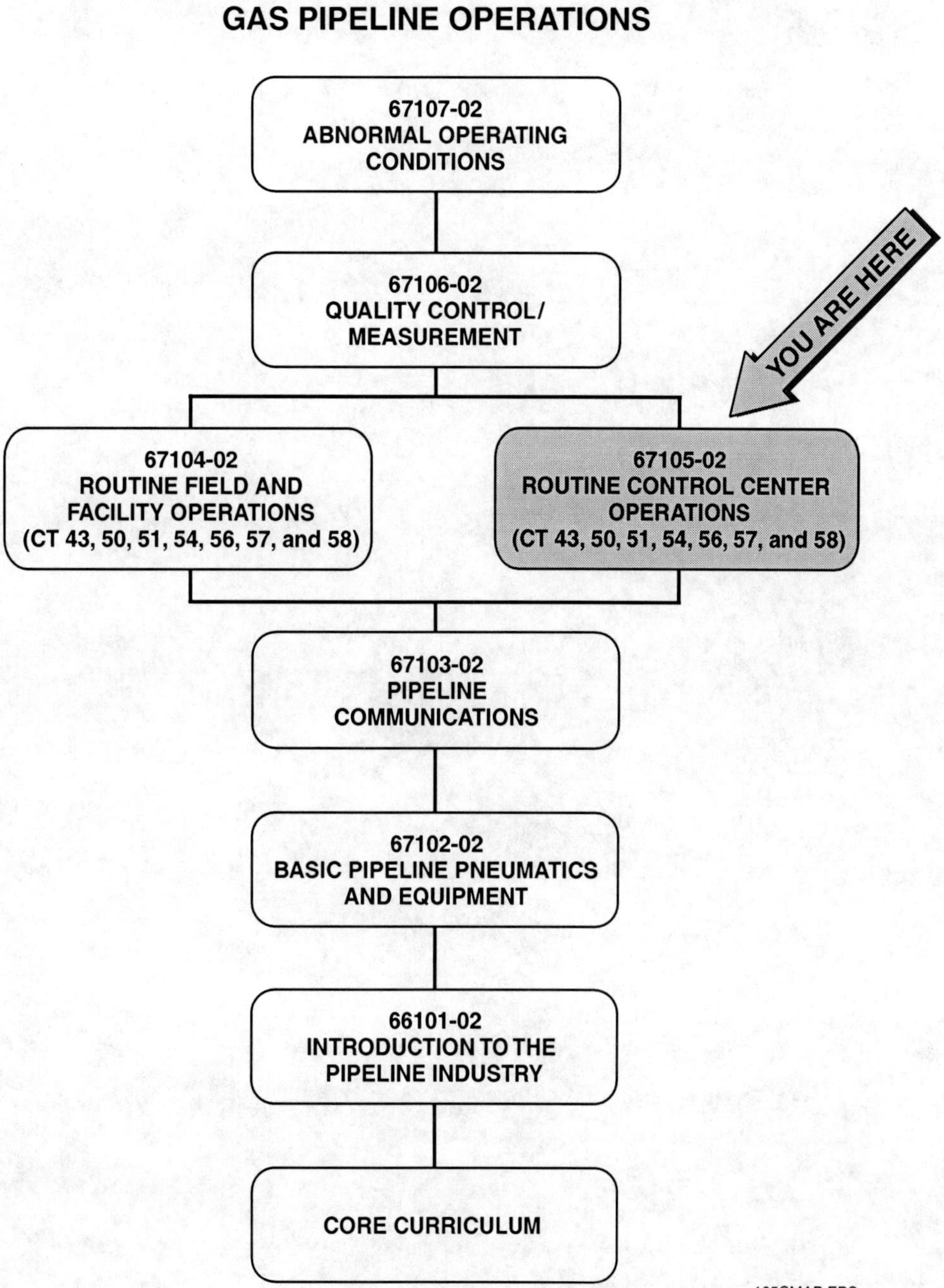

105CMAP.EPS

MODULE 67105-02 CONTENTS

MODULE 67105-02 CONTENTS (Continued)

Figures

Table

MODULE 67105-02

Routine Control Center Operations (CT 43, 50, 51, 54, 56, 57, and 58)

Objectives

When you have completed this module, you will be able to do the following:

1. Perform SCADA operations and explain the concepts and applications of the SCADA system.
2. Perform pipeline system monitoring and pipeline station monitoring with the SCADA system (CT 43.3).
3. Document pipeline activities with the SCADA system.
4. Explain manifold operations (CT 43.4) and compressor operations.
5. Monitor and respond to alarms and test remotely controlled shutdown devices (RCSDs) (CT 54).
6. Describe the general procedures for performing system startup (CT 43.1) and system shut in/shutdown (CT 43.2).
7. Describe the general procedures for purging gas from pipelines (CT 50), purging air from pipelines (CT 51), and monitoring purging during maintenance.
8. Perform pigging operations.
9. Describe the general procedures involved in uprating the pipeline's maximum allowable operating pressure (MAOP) (CT 56).
10. Operate odorant equipment (CT 57) and monitor odorant level if applicable (CT 58).

Prerequisites

Before you begin this module, it is recommended that you successfully complete the following modules: Core Curriculum; Pipeline Core, Module 66101-02; Gas Pipeline Operations, Modules 67102-02 and 67103-02.

Required Trainee Materials

1. Pencil and paper
2. Appropriate personal protective equipment

1.0.0 ◆ INTRODUCTION

The **control center** (see *Figure 1*) is at the heart of pipeline operations. Information about the pipeline's operating equipment and parameters flows into the control center, where personnel using computers monitor it. Control center personnel also monitor signals transmitted to pipeline equipment and communicate with field personnel when necessary. This module covers the general operation of a pipeline's control center. In addition, it covers some of the tasks performed on the pipeline that relate to control center operations. Even though control center personnel may not have to perform some of the tasks covered in this module, they must have a good understanding of the steps involved in these tasks to do their jobs effectively.

2.0.0 ◆ THE SCADA SYSTEM

The Supervisory Control and Data Acquisition (SCADA) computer system allows personnel to monitor alarms and operating data and remotely control equipment, such as valves and compressors, that may be hundreds of miles away from the control center. Because pipeline operations take place over long distances, control center personnel also use other communications systems, such as telephone and radio, to communicate with facility stations and field personnel.

105F01.EPS

Figure 1 ◆ Typical control center.

The SCADA system continuously monitors the volume in the line and provides line balance reports and over-and-short calculations. Some SCADA systems offer multiple screens so that pertinent facts at any location are instantly available. In addition, the SCADA system can display information graphically so that it is more easily understood. If a change occurs, such as a valve opening or closing, the shape representing that valve on-screen changes as well. Pipeline variables such as pressure may be shown graphically or numerically. Most SCADA systems have multiple screens so that control center personnel can select relevant ones.

2.1.0 System Devices

SCADA system devices measure, transmit, organize, analyze, and store data. These devices include field devices, **remote terminal units (RTUs), programmable logic controllers (PLCs),** a communications network, a host (or master) computer system, and operating software. See *Figure 2* for a schematic overview of a pipeline monitoring system.

2.2.0 Field Devices

Field devices are placed along the pipeline at strategic locations to measure the composition of the gas and to monitor pressure and flow. Other sensors are placed in critical locations, such as bearings where parameters cannot exceed set limits without causing serious damage. Alarms are included in field devices. Data is transmitted to the control center where the host computer displays the ongoing readings on meters and records the data for calculation and verification.

2.2.1 Measuring Gas Composition

Gas chromatographs are used to measure the composition of gas. The chromatograph separates the volatile organic compounds that make up the gas. These compounds pass through a column filled with a special material where their molecules are separated for analysis. Results of the analysis are printed on a graph.

Natural gas is measured in volume units. The common unit is a cubic foot of natural gas at 60°F and an atmospheric pressure of 14.7 pounds per square inch (psi).

British thermal units (Btus) measure the energy generated from the burning of a unit volume of natural gas. One Btu is the heat needed to raise the temperature of one pound of water from 60°F to 61°F at normal atmospheric pressure.

A cubic foot of natural gas gives off 1,000 Btus, on average, but the range is 500 to 1,500 Btus. The measurement varies because natural gas supplies vary in the energy gases they contain. The higher the amount of combustible gases in a natural gas, the higher the Btu.

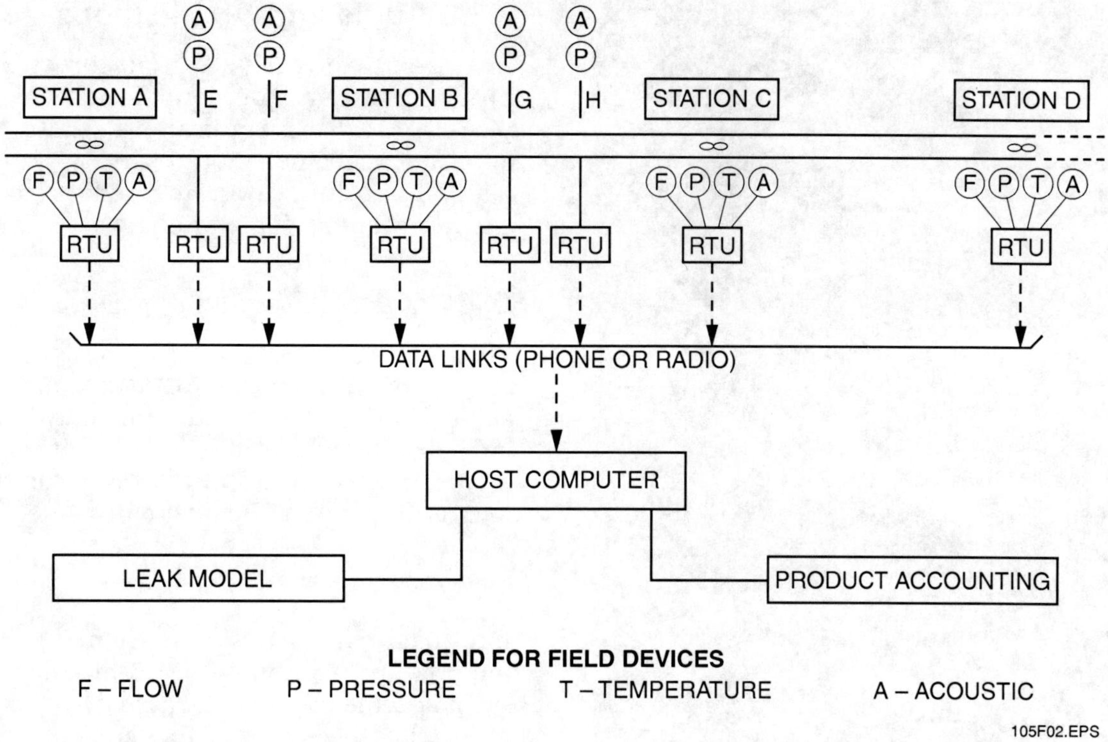

Figure 2 ◆ Schematic overview of a pipeline monitoring system.

2.2.2 Measuring Flow

Flow is an important variable at custody transfer stations. Flow meters installed in the pipeline system measure and record the flow rate. Many companies require line balance to be checked weekly or monthly. The volume of gas moving through the system directly relates to the rate of flow. Flow data is sent to the control center where the volume is calculated and the data recorded for future use.

2.2.3 Measuring Pressure

Pressure is important to monitor because it directly affects many other aspects of pipeline operation. Modern pressure sensors can be programmed to read all of the variables that can affect pressure, such as temperature and volume. Ongoing pressure readings are displayed in the control center and recorded for calculation and verification. Alarms are included on pressure circuits to alert control center personnel to problems.

2.3.0 RTUs and PLCs

RTUs and PLCs scan the connected field inputs for data. The host computer in the control center processes the data to determine if there is an alarm condition.

RTUs may do some data processing, but most record and file data until SCADA scans the unit. The RTU downloads the data to the SCADA host computer for processing.

PLCs (see *Figure 3*) are small industrial computers containing a program that operates a loop with inputs and outputs. These computers scan the loops, gather the data, and initiate appropriate outputs. The PLC can monitor individual circuits, differentiate between the normal and alarm states, and signal the host computer when any circuits enter the alarm mode. Programs are set up to discern analog, digital, and pulse signals.

2.4.0 Communications

SCADA operating systems generally have two communications levels: client and data server. The client level provides an interface between personnel and their consoles. The data server provides an interface that allows data to flow to the host computer from the PLCs and RTUs.

2.5.0 Host Computer and Software

The host computer, located in the control center, is the heart of the SCADA system. The host computer consists of a mainframe and desktop units linked in a network. Most host computers employ off-the-shelf operating systems that interface with custom programs or generic software.

Figure 3 ◆ Programmable logic controller.

2.6.0 Data Interpretation

The host computer continuously receives analog and digital data on temperature, pressure, and flow. Sensors that measure pipeline parameters, such as pressure and temperature, transmit analog signals that are numerically displayed and may be trended for display on a graph.

PLCs exchange pressure signals with the SCADA computer. The computer combines the data into a series of signals and organizes it into a graphic display. The display represents the pressure levels at all the points along the pipeline. At the same time, the pressure levels are kept in a log that can be accessed anytime.

Normally, the signals from the host computer and the PLC match each other. Sometimes, however, they do not. For example, the host computer may send a signal representing the setpoint for pressure to the PLC. The pressure sensor records the actual pressure on the line. The PLC compares the two signals. The variance between them is too great, so the PLC alerts the host computer, which initiates an alarm. Control center personnel then respond in accordance with company policy.

2.7.0 Tools and Applications

In order to effectively monitor and control pipeline systems, personnel must understand SCADA tools and applications. These tools and applications include graphic displays, event summaries, trending programs, redundancies, and communications verifications.

2.7.1 Graphic Displays

Much of the information that is sent to the host computer is graphically displayed on screens. At some companies, the entire pipeline system may be viewed graphically. A screen capture provides a hard copy of a graphic display for reference, logs, and required documentation.

2.7.2 Event Summaries

Information in the SCADA system is managed by logging and archiving electronic files. Logging saves specified files on the hard drive for the short term. Once the log is full, files are transferred to an archive file. Archive files store information permanently. Data is time stamped so that information can be retrieved easily.

2.7.3 Trending Programs

Some SCADA systems include programs that present data in multiple displays. Control center personnel can see readouts of pressure, temperature, flow, and other parameters as they occur. At the same time, the computer displays the data through a real-time graph. The user can define events or parameters to create historical trending charts from archived data. By observing the trending graphic, control center personnel can see the conditions and changes on the pipeline.

2.7.4 Redundancies

Most SCADA systems are designed with redundancy, or backup, systems to protect the pipeline. Elements in the electronic equipment are also intentionally duplicated so that if one channel fails, another can keep the pipeline operations running. To get around transmission errors, some signals are repeated.

SCADA systems use the main communications system during normal operation, while backup systems are standing by. If the main computer goes down, a second communications system goes online instantly.

2.7.5 Communications Verifications

The SCADA host computer scans all communication circuits at specified times to verify that all circuits are operating. RTUs receive requests for data at specified intervals. If the primary communications channel fails, backup communications come online.

3.0.0 ◆ MONITORING PIPELINE SYSTEMS AND STATIONS

Control center personnel monitor important factors such as pressure, flow rate, and status changes on the pipeline and at facility stations. The SCADA system reports and displays deviations quickly. To identify and respond quickly to deviations, control center personnel should be familiar with the physical pipeline system. They should be aware of operating conditions, equipment being operated upstream and downstream, and weather conditions.

3.1.0 Monitoring Pressure

Pressure is the force that a gas exerts uniformly in all directions within a vessel, pipeline, hole in the ground, or other containment area. Pressure is measured as force exerted per unit of area, such as pounds per square inch.

Virtually every action on the pipeline depends upon keeping the pressure properly controlled. If the pressure goes too high, the pipeline can rupture. Certain equipment, metering devices, and terminal installations often must be protected from main-line pressure. If the pressure falls too low, service to customers will be affected.

Many companies require control center personnel to begin their shift by checking and verifying pipeline pressure, compressor pressure, delivery pressure, and relief setpoints. Personnel observe the readings on the meters and on the appropriate monitor screen. Control center personnel should be familiar with the normal operating pressures for the pipeline under similar conditions and verify that current readings are within normal range. Knowing what is normal helps personnel recognize what is *not* normal. Personnel should continuously monitor pressures using the SCADA system, pressure gauges, or meters, looking for any changes. Computer screens and analog readings provide an ongoing display of pressure in several formats. Personnel should note changes in pressure before alarms sound.

3.1.1 Deviations in Pressure

Deviations in pressure can occur at any time, and control center personnel should identify them quickly so that appropriate actions may be taken. Pressure deviations may not be serious enough to require anything except close monitoring. More serious deviations may require that control center personnel take actions in accordance with company policy.

Pressure detection devices, located at all critical locations on the pipeline, send signals that update the computers in the control center every few seconds. An alarm sounds only when the line pressure exceeds the setpoint.

3.1.2 Responding to an Alarm

If an alarm sounds and the pressure remains too high or low, control center personnel contact the appropriate personnel to determine the cause. Each possible cause should be investigated.

Changes in pressure may be caused by compressors that have been started or stopped at other stations or by valves unexpectedly opening or closing. Control center personnel should determine the cause and confirm it before initiating any action. Generally, if the situation cannot be addressed by the control center, field personnel may be instructed to investigate the cause.

3.2.0 Maintaining the System within Set Parameters

For safety and efficiency, pipeline facilities and systems are designed to operate within certain limits or parameters. The maximum allowable operating pressure (MAOP) is established for each pipeline system or segment. Safety devices and related alarms alert control center personnel when the system strays from within the established operating pressure limits.

3.2.1 Knowing System Parameters

Qualified control center personnel must be completely familiar with the entire pipeline system and its parts. They should be able to look at the pressure meters, gauges, and rate of flow and correctly identify conditions on the line at all times. In addition, they must be completely familiar with the company's policies and procedures. This knowledge enables control center personnel to quickly and correctly respond when any problem occurs.

3.2.2 Operating within Limits

The MAOP for each pipeline system is established during the design phase of the system. The MAOP will not change unless there is a physical change in the pipeline system or the pipeline is **derated** because of the effects of corrosion. Normally, pressure readings steadily vary from 1 to 2 psi; unusual variations in pressure could indicate a serious problem.

If pressure begins to change at some point in the line, the SCADA system shows it within seconds. Control center personnel see the pressure fluctuation by observing the analog or digital meters and monitor screens. A rise in pressure at some point could indicate an unintended valve closure. Control center personnel must investigate every possible cause and quickly return the pipeline to normal operating limits.

3.3.0 Station Monitoring

Personnel also monitor all **compressor stations** and related equipment from the control center. The SCADA system receives all operating information from individual stations and shows conditions in real time. In certain situations, control center personnel must switch control from the control center to an individual compressor station. This is usually done for routine maintenance, but may also be done in response to an emergency situation. The SCADA system continues to receive information from the station so that there is no break in data recording. To ensure safety, control center personnel must follow company procedures when switching control to stations.

Control center personnel are responsible for monitoring pressure at all critical points in each station. Following are the pressure points that should be monitored:

- Suction pressure and discharge pressure
- Line pressure
- Differential pressure

3.3.1 Suction Pressure and Discharge Pressure

To work efficiently, compressors must have an ample supply of gas. Suction pressure is the amount of pressure generated on the upstream side of the compressor. Discharge pressure is the amount of pressure generated on the downstream side of the compressor and supplied to the pipeline. Pressure transmitters (see *Figure 4*) help regulate the proper amount of suction and discharge and the proper flow of gas.

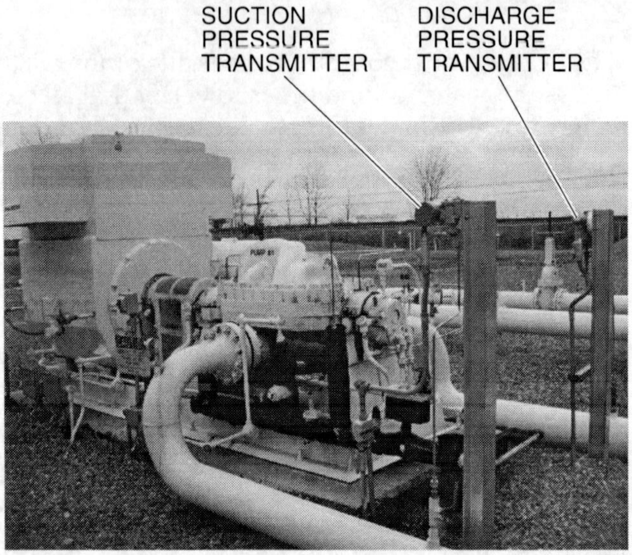

Figure 4 ◆ Suction and discharge pressure transmitters.

The SCADA system compares suction pressure against discharge pressure, and the control valve opens or closes as necessary to maintain the proper amount of suction. The suction is constantly updated on meters in the control center and on the SCADA monitor screens. When the compressor starts, the discharge pressure and flow rate increase and suction decreases. Adjustments may be made to keep suction and discharge pressure within an acceptable range.

3.3.2 Line Pressure

Line pressure, the pressure on the main pipeline, is an indicator of system integrity. A line pressure transmitter (see *Figure 5*) takes pressure measurements and relays them to the control center.

Figure 5 ◆ Line pressure transmitter.

3.3.3 Differential Pressure

Differential pressure is the difference between the pressure at two points in a closed system, such as the upstream and downstream sides of a gate valve. Differential pressure at various locations should be monitored so that the pressure change does not damage piping and equipment.

3.4.0 Monitoring the Pipeline Right-of-Way

Because pipelines cover so many miles, companies must use several methods to survey the pipeline (*49 CFR 192.705*). These methods include walking, driving, and flying. In addition, video equipment allows personnel to monitor certain areas remotely. Gas pipelines must be patrolled regularly (see *Table 1*). Inspection personnel check the pipeline right-of-way (ROW) for any conditions affecting the integrity of the pipeline. Personnel check for a variety of conditions along the pipeline ROW. Following are examples of some of the conditions they look for:

- Construction work
- Soil erosion or growth of vegetation
- Missing or damaged pipeline markers
- Encroachment
- Signs of vandalism

Table 1 Maximum Interval between Patrols

Class Location of Line	Maximum Interval between Patrols	
	At Highway and Railroad Crossings	At All Other Places
1 and 2	7½ months, but at least twice each calendar year	15 months, but at least once each calendar year
3	4½ months, but at least four times each calendar year	7½ months, but at least twice each calendar year
4	4½ months, but at least four times each calendar year	4½ months, but at least four times each calendar year

3.4.1 Monitoring by Aircraft

Aerial patrols, which are the most common form of monitoring, are performed on a schedule set by the pipeline company. Using current maps of the pipeline, pilots fly slowly at a safe altitude to allow for a good view. They radio their reports to the control center. Most companies require aerial patrols to keep a written record of their observations. In some cases field personnel are sent to investigate an aerial report and complete a follow-up report. Pilots may be required to review any follow-up reports and compare them with the original patrol log.

3.4.2 Monitoring by Vehicle and Foot Patrol

Although aerial patrols are more commonly used to inspect the pipeline, experienced ground patrols, called **linewalkers,** may be required to inspect areas where low-level flying is prohibited. In addition, linewalkers may be required when visibility is low or when bad weather grounds aircraft. In some cases, linewalkers are sent to investigate an aerial report more closely. Linewalkers may inspect the line on foot or may drive along areas that are accessible by car.

3.4.3 Monitoring by Remote Video

Some facilities provide 24-hour security, as well as video cameras and monitors (see *Figure 6*). Being able to continuously see all parts of a facility is important. Video cameras and monitors can reveal equipment failure, as well as break-ins and other security problems. Surveillance cameras can also focus on critical equipment and locations so that security and field personnel can monitor these areas for problems.

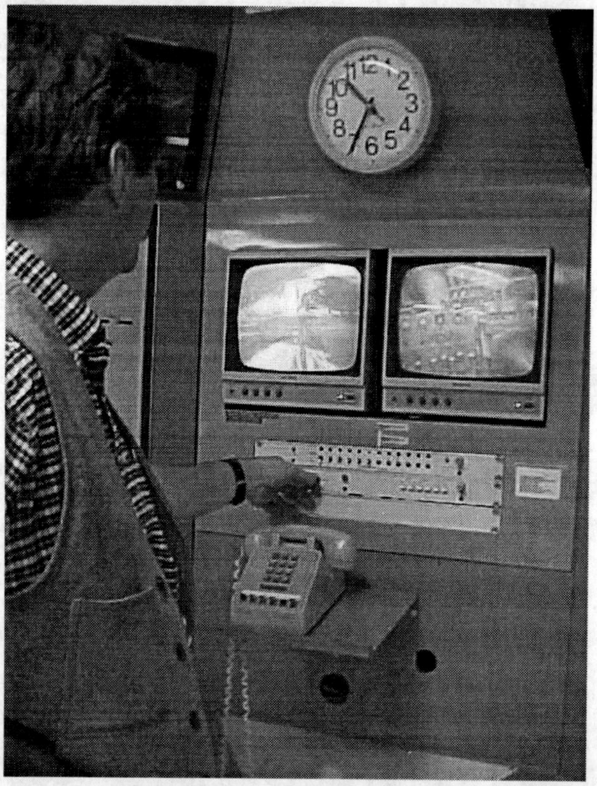

105F06.EPS

Figure 6 ◆ Typical video monitors.

3.5.0 Monitoring Weather Conditions

Many pipeline operations depend on moderate weather. A tornado, hurricane, or blizzard can threaten personnel and the pipeline itself. High winds, snow, ice, and cold temperatures may also hinder operations. Sometimes the approach of an electrical storm signals the danger of deadly lightning and wind.

The weather must be monitored to protect the safety of employees working on the pipeline. In addition, company procedures should be followed for properly recording all weather changes.

The sooner that personnel become aware of a breaking weather situation, the more likely it is that they can take action before problems become critical. To monitor weather changes effectively, personnel sometimes rely on weather-reporting Web sites. Many stations use small weather monitors to gauge local conditions. Following are examples of some conditions that are monitored:

- Temperature
- Wind direction and speed (with high-wind alarm)
- Wind chill factor (with low-wind-chill alarm)
- Daily rainfall
- Barometric pressure

3.6.0 Documenting Pipeline Activities

Everything that happens on the pipeline is documented for company records. The SCADA system keeps records of all pressure, temperature, flow, and any other pertinent readings, and personnel can retrieve them at any time. Archived files are kept permanently.

Companies also investigate and document **abnormal operating conditions (AOCs)** for their own records and to provide documentation required by state and federal regulations. Required documentation includes the following paperwork:

- Incident reports
- Event notification forms
- Incident investigation and analysis reports
- Control center logs
- Station logs
- Records of recommendations for corrective actions and training, if applicable

3.7.0 Responding to Abnormal Operating Conditions

Pipeline personnel must respond to AOCs, such as a gas leak, equipment malfunction, or facility damage quickly but not hastily. It is important to follow established company procedures so that each problem is confirmed, investigated, and addressed safely and correctly. Personnel confirm alarms to ensure that an alarm is not a false alarm. They investigate problems to determine what is wrong and what needs to be done to correct it. Depending on the situation, steps may have to be taken immediately to secure the area and ensure the safety of personnel. Other situations may require that repairs be scheduled and necessary work permits issued. In general, personnel follow these steps when responding to an AOC:

Step 1 Evaluate the situation.

Step 2 Notify affected personnel.

Step 3 Carry out company safety procedures to protect personnel, property, and the environment.

Step 4 Make necessary repairs or report that repairs are required.

Step 5 Submit the required reports.

Review Questions

Sections 2.0.0–3.0.0

1. _____ are used to measure the composition of gas.

a. Gas chromatographs
b. RTUs
c. PLCs
d. BTUs

2. _____ are small industrial computers containing a program that operates a loop with inputs and outputs.

a. RTUs
b. PLCs
c. BTUs
d. SCADAs

3. _____ is the force that a gas exerts uniformly in all directions within a vessel, pipeline, hole in the ground, or other containment area.

a. Gravity
b. Flow rate
c. Discharge
d. Pressure

4. The MAOP for each pipeline system is established during the _____ phase of the system.

a. design
b. construction
c. testing
d. startup

5. _____ pressure is the amount of pressure generated on the upstream side of the compressor.

a. Outward
b. Inward
c. Discharge
d. Suction

4.0.0 ◆ PERFORMING MANIFOLD OPERATIONS

A valve is a mechanical device that controls the flow of gas through a pipeline. Some valves are automatic safety devices. Sometimes multiple valves are grouped into a **manifold,** or header, and are remotely controlled.

4.1.0 Operating Valves by Remote Control

When valves are wired to RTUs, they can receive commands from the control center, which may be hundreds of miles away. RTUs send signals, which report the status of the valves, to the control center. They also receive commands from the control center to open and close the valves. Valve control units on the control center monitors show the position of each valve. Control center personnel can see at a glance which valves should be opened or closed or when valves do not respond to commands. When control center personnel open and close valves in the correct sequence, the gas is safely directed through the proper valves and piping.

4.2.0 Types of Valves

Several types of valves are required for safe and efficient operation of the pipeline. These valves include relief valves, block valves, manually controlled valves, flow control valves, and check valves.

4.2.1 Relief Valves

Relief valves allow pressure to be safely relieved at critical points. Many companies require control center personnel to begin their duty shift by checking and verifying pressure levels and setpoints on relief valves. Personnel check the pressure by reading analog or digital meters (see *Figure 7*) and the appropriate monitor screens. They can adjust the setpoints as necessary by remote control.

Figure 7 ◆ Analog and digital meters.

4.2.2 Block Valves

Block valves are the main valves controlling gas coming into the station. These valves may be controlled manually or by remote control and are operated either fully opened or fully closed. Personnel close this type of valve to shut in, or isolate, a **pipeline segment** for maintenance or repairs.

4.2.3 Manually Operated Valves

Although many valves are operated by remote control, some valves must be turned by hand. Manually operated valves are common on meter runs (see *Figure 8*).

4.2.4 Flow Control Valves

Flow control valves are installed at custody transfer stations, which is where a change of ownership or a change of responsibility for the natural gas in a pipeline takes place.

These valves have a flow setpoint with a pressure override. The override prevents an excessive buildup of pressure as gas moves through the custody transfer station. Operations personnel enter the setpoints for each valve so that a suitable operating pressure is maintained.

Figure 8 ◆ Manually operated valves.

Flow control valves are also installed in systems where constant-speed, main-line compressors are used. Such valves are necessary to control, or throttle, pressure to the pipeline from the compressor's discharge. These valves are normally operated in a partially opened setting. Moving the valve to a partially closed setting prevents pressure and flow from exceeding the pressure setpoints. Operations personnel calculate the setpoints and adjust the valves from the control center to maintain correct operating pressure.

4.2.5 Check Valves

Check valves allow air to flow in one direction and block it from flowing upstream, or back through the pipeline. This type of valve has a suction port at one end of the valve body, a discharge port at the other end of the valve body, and a movable part, or seal, that is activated by spring pressure. The seal might be a flapper, plunger, ball, or poppet. When the pressure at the suction port is higher than the pressure of the spring that holds the seal in place, the seal moves off its seat. Air then flows freely through the valve. When air tries to return through the discharge port, it pushes the seal back onto its seat, and airflow through the valve is blocked.

4.3.0 Valve Alignment

Control center personnel must set up and verify **valve alignment** (see *Figure 9*) for every change in operations. They determine the safest and most efficient way to move the gas from point A to point B by studying piping layouts and selecting the correct valve sequence. The valve sequence may be marked on a printed chart or simply listed on a piece of paper. Valve alignment involves opening and closing valves in the right sequence, or order. Although some control center personnel may learn the proper valve sequence by heart, they must always refer to their valve alignment chart before proceeding.

Some companies require personnel to follow a set procedure to confirm that valves are in the correct sequence before any operational change. For example, the company may require control center personnel to record the position of each valve in a log before proceeding with any operational changes.

Control center personnel may be required to contact field personnel to verify that the valve alignment is set properly. There is always the possibility that one or more valves won't respond to a signal from the control center. In addition, problems with other pipeline equipment may also prevent proper alignment. If there is a problem, field

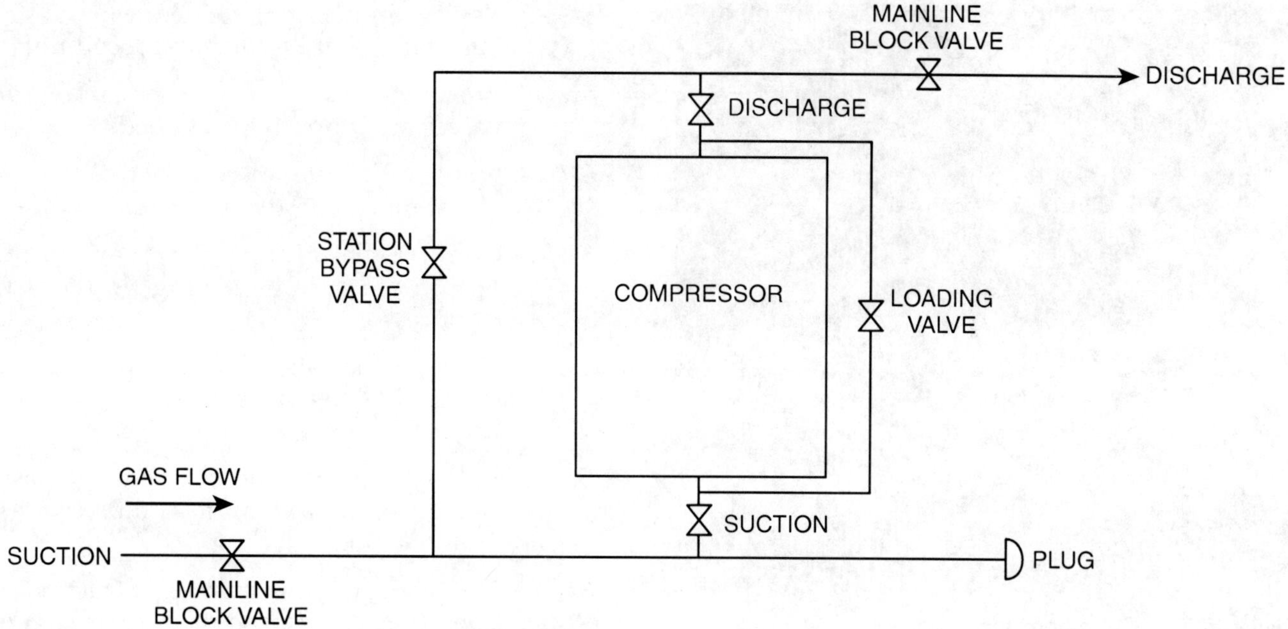

Figure 9 ◆ Valve alignment.

personnel must inspect the valves to ensure that they are all properly set and that pipeline equipment is operating.

4.3.1 Flow Path

Valves must be properly aligned and operating in the right sequence to ensure a clear flow path through the line. The flow path is adversely affected if the valves are not properly aligned. In addition, malfunctioning or blocked valves will also negatively affect the flow path. If there is a problem with a valve, the control center notifies the appropriate maintenance supervisor.

4.3.2 Valve Interlocking

Valve interlocking is performed by a computer program that operates a number of valves in a particular sequence. For example, if there is a change from normal conditions to an alarm condition, the program starts an automatic control sequence that closes the valves automatically at a command from the control center. These valves are interlocked with a main valve at another location downstream and cannot close until the main valve closes.

4.4.0 Custody Transfer

Each pipeline company has a standard procedure that must be followed when transferring custody of gas. An understanding of the pipeline's operating parameters is essential. All equipment at the custody transfer station must be carefully monitored to ensure that the right amount of gas is transferred at the correct pressure. Flow control valves installed at the custody transfer station are fitted with a pressure-override device, which prevents an excessive buildup of pressure as the gas moves from one custody area to another.

5.0.0 ◆ OPERATING COMPRESSORS

Natural gas is sent through a pipeline at high pressures. As the gas moves through the pipeline, friction reduces its speed and pressure drops. To maintain the necessary pressure, the gas must be recompressed periodically as it moves through the pipeline. This is why compressor stations are installed approximately every 100 miles along the pipeline. Often, several compressors are at one station.

Some compressors use a small part of the gas that flows through the pipeline to power the engine that drives the compressor. Some compressors are electrically powered by turbine engines. An example of a field compressor is shown in *Figure 10.* A natural gas compressor is shown in *Figure 11.*

Several different types of compressors are available for use on the gas pipeline. Therefore, it is important to read and follow the manufacturer's recommendations for installation, startup, shutdown, and maintenance procedures. Documentation for these procedures should be available at the pipeline facility for easy reference if needed.

105F10.EPS

Figure 10 ◆ Field compressor.

105F11.EPS

Figure 11 ◆ Natural gas compressor.

5.1.0 Compressor Startup

The procedures and sequence for starting up a compressor vary from one company to another. In addition, each company has its own notification and reporting requirements. In general, the following steps are taken to start a compressor:

Step 1 Carry out any pre-startup procedures as recommended by the manufacturer or required by company policy.

Step 2 Review and verify knowledge of the starting unit and of the startup procedure.

Step 3 Ensure that the compressor and station valves are properly positioned.

Step 4 Notify all affected personnel that the compressor is about to be started.

Step 5 Confirm that all operating variables are within the safe operating limits of the compressor.

Step 6 Confirm that all safety alarms and shutdown devices are set and are working properly.

Step 7 Initiate the startup according to the posted startup procedure for the compressor(s) at your location.

Step 8 Follow the manufacturer's instructions for operating the compressor after it has been started.

The manufacturer's instructions include information regarding monitoring. The following items are monitored during operation of the compressor:

- Suction and discharge pressures
- Operating temperatures, pressures, and speeds
- Unusual noises or vibrations
- Leaks
- Surges

5.2.0 Compressor Shutdown

Compressors and drivers may have to be shut down for regular maintenance or in response to a problem. This procedure must be performed carefully, safely, and in accordance with the manufacturer's specifications. In general, the following steps should be taken to shut down a compressor:

Step 1 Slow down the compressor to minimum speed.

Step 2 Unload the compressor and open the bypass valve.

Step 3 Allow the compressor to cool down.

Step 4 Shut off the fuel supply.

Step 5 Close the suction and discharge valves.

Step 6 **Blowdown** the compressor.

The amount of time that the compressor may be left depressurized depends on a number of factors, including the condition of the compressor, the need to make repairs, and the availability of backup compressors. Generally, the maintenance supervisor decides whether the compressor should remain out of service.

5.3.0 Abnormal Operating Conditions

Several AOCs are associated with startup and shutdown of a compressor. When AOCs occur, personnel must follow company policy for identifying the problem and notifying the appropriate personnel. In addition, personnel must conduct any investigations or submit reports required by law or company policy. Following are two common AOCs:

- *Gas leak*—A hazardous gas alarm will sound. Personnel may be required to leave the area. The gas leak must be repaired and the air checked with a **combustible gas indicator (CGI)** to ensure safety.
- *Excessive vibration or noise*—Too much vibration or noise may indicate that equipment is malfunctioning or about to fail. The compressor may have to be shut down while personnel investigate the cause.

6.0.0 ◆ TESTING REMOTELY CONTROLLED SHUTDOWN DEVICES

Remotely controlled shutdown devices (RCSDs) are electronic devices that are used to shut down the compressor driver. Each RCSD must be inspected and tested at least once each calendar year to ensure that it is working properly (*49 CFR 192.731*).

You must know how to do the following tasks:

- Ensure that each RCSD is receiving electrical power.
- Test high-temperature shutdown devices.
- Test high-discharge-temperature shutdown devices.
- Test high-liquid-level shutdown devices.
- Test excessive-vibration shutdown devices.
- Test engine-overspeed shutdown devices.
- Test inadequate-lubrication shutdown devices.
- Test high- and low-suction-pressure shutdown devices.
- Properly document the maintenance inspection.

To inspect and test RCSDs, personnel must have a thorough knowledge of the following:

- The company safety procedures
- The company procedures for responding to emergencies
- The shutdown and safety devices being inspected
- The procedures for performing inspections
- The **piping and instrumentation diagrams (P&IDs)** associated with the equipment being inspected

6.1.0 Testing Procedure

Before testing the RCSD, employees must review the testing procedure. This review is important for the integrity of the system, personal safety, the safety of others, and proper inspection. Personnel must always wear the appropriate personal protective equipment (PPE) during the test.

Before the test begins, all affected personnel and the control center must be told that a station is preparing to test an RCSD. The control center must also be notified when the test is completed. The proper sequence for conducting the test follows:

Step 1 Activate the RCSD.

Step 2 Inspect the RCSD to ensure that all safety, shutdown, alarm, and isolation equipment is working properly, as described on the P&IDs.

Step 3 Make any necessary repairs, or report that repairs are needed.

Step 4 When the RCSD inspection is completed and the RCSD is reset and ready to return to service, notify affected personnel.

Step 5 Submit the required reports.

6.2.0 Abnormal Operating Conditions

During an RCSD test, there may be a gas leak. Personnel should isolate the source of the leak if it is safe to do so. In addition, personnel must carry out established company safety procedures to protect people, property, and the environment. Once conditions are safe, the leak may be repaired. During repairs, the air must be continuously monitored for hazardous gas.

Review Questions

Sections 4.0.0–6.0.0

1. _____ valves are installed in systems where constant-speed, main-line compressors are used.
   a. Block
   b. Check
   c. Flow control
   d. Relief

2. Valve _____ involves opening and closing valves in the right sequence, or order.
   a. consignment
   b. alignment
   c. refinement
   d. assignment

3. To keep gas compressed and moving through the pipeline, compressor stations are necessary approximately every _____ miles along the pipeline.
   a. 50
   b. 100
   c. 200
   d. 500

4. The first step in shutting down a compressor is to _____.
   a. allow the compressor to cool down
   b. shut off the fuel supply
   c. close the suction and discharge valves
   d. slow down the compressor to minimum speed

5. Remotely controlled shutdown devices are electronic devices that are used to _____.
   a. shut down the compressor driver
   b. blowdown the compressor
   c. exhaust discharge pressure
   d. fine-tune suction pressure

7.0.0 ◆ MONITORING SAFETY DEVICES

Safety devices are an essential part of monitoring pipeline operations. Some devices are designed to detect changes in pressure. Other devices are designed to detect hazardous atmospheres. Following are the types of safety devices found in pipeline operations:

- Pressure relief systems
- High-pressure shutdown devices
- Hazardous atmosphere devices
- Detection systems
- Gas detection devices
- Smoke, heat, and flame detection devices

When a safety device alarm goes off, it must never be ignored or turned off until the situation has been addressed. Generally, the control center confirms the alarm and notifies field personnel. Field personnel may be directed to investigate alarms to ensure that alarm equipment is working properly.

7.1.0 Pressure Relief Systems

Pressure relief systems are designed to prevent pressure from exceeding the MAOP. To help keep the pipeline pressure at the correct MAOP, control center personnel calculate relief valve setpoints. Relief valves are designed to open instantly and close slowly and smoothly so that excess pressure is relieved while keeping the pipeline in balance.

7.2.0 High-Pressure Shutdown Devices

High-pressure shutdown devices activate when pipeline pressure exceeds established setpoints. A main-line compressor shuts down as a result of a high-pressure discharge when setpoints are exceeded. Such an event could indicate a problem in the pipeline. In such cases, an alarm will sound in the control center. Control center personnel must notify the appropriate personnel immediately when an alarm sounds. A trained and qualified employee investigates equipment in accordance with company policy.

7.3.0 Hazardous Atmosphere Devices

At some compressor stations, **lower explosive limit (LEL),** thermal, and **ultraviolet (UV) detectors** continuously sample the air for hazardous atmosphere, high heat, and flames. When the detectors sense an abnormal condition, they activate warning lights and alarms.

7.4.0 Detection Systems

Detection systems (see *Figure 12*) detect fire and unsafe levels of hazardous gas. These systems also protect field equipment and facilities. These systems sound alarms, discharge **fire suppression material,** and shut down equipment. Fire detection and suppression systems protect by doing the following:

- Monitoring hydrocarbon gas levels
- Detecting smoke, flames, and excessive temperatures

105F12.EPS

Figure 12 ◆ Fixed gas detection system.

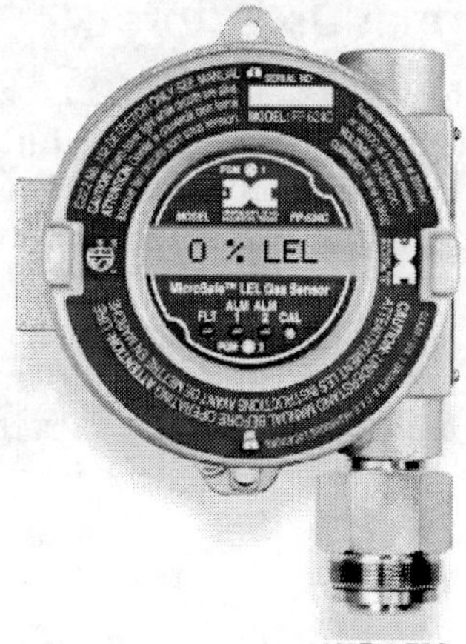

105F13.EPS

Figure 13 ◆ Combustible gas sensor.

- Alerting personnel to a fire or hazardous atmosphere
- Activating the emergency ventilation system
- Isolating hydrocarbon sources
- Automatically shutting down equipment
- Automatically blowing down large-volume natural gas systems
- Automatically discharging fire suppression material

In addition, personnel are able to manually discharge fire suppression material by using pull switches. They may do this if the automatic system fails to work.

7.5.0 Gas Detection Devices

Combustible gas detection devices (see *Figure 13*) provide early warning of hazardous atmospheres so that personnel can take appropriate safety steps. By law, the devices must signal an alarm when gas levels reach 25% of the LEL. Most alarms are set to warn at 20% of the LEL. When gas levels reach 60%, equipment shuts down and fire suppression material is dumped. If a monitor is reading close to the LEL and suddenly drops to 0%, this may indicate that the area has filled with gas. No alarm can be reset until trained personnel wearing **self-contained breathing apparatus** investigate the condition.

7.6.0 Smoke Detectors

Ionization detectors (see *Figure 14*) sense combustion even before smoke or flames appear. Ionization detectors work well for detecting fire in electrical conduit cable trays. They can detect smoke in facility buildings, crawl spaces, generator rooms, control rooms, and areas containing a lot of electrical equipment.

105F14.EPS

Figure 14 ◆ Ionization detector.

7.7.0 Flame Detectors

UV detectors sense fires almost instantly, even when the flames have barely started. In addition, these detectors can distinguish between the UV rays produced by fire and the rays produced by natural light.

That same sensitivity, however, can also cause false alarms. To prevent false alarms, at least two different, overlapping sensors (**voted systems**) must sound an alarm before personnel can act. Causes of false alarms include welding, electrical arc sources, camera flashes, and X-rays or gamma rays.

UV systems work best when nothing gets between the sensor and the fire. Fires that produce heavy smoke make it harder for these systems to detect a fire.

7.8.0 Thermal Detectors

Thermal detectors (see *Figure 15*) react to rapidly rising temperatures by closing controls that are normally open. Thermal detectors are used with graduated temperature settings. This spread of settings allows for special applications. Thermal detectors may be adjusted for a particular setpoint. An early warning sensor closes the controls and sounds an alarm before the setpoint is fully reached.

105F15.EPS

Figure 15 ◆ Thermal detector.

8.0.0 ◆ PERFORMING SYSTEM STARTUP AND SEGMENT SHUT IN

In gas pipeline operations, a full system startup is performed only on a newly installed system. In addition, the entire pipeline system is almost never shut down. Usually, a pipeline segment is shut in, or isolated, for maintenance or repairs. Shutting in the segment allows the gas to keep flowing while maintenance or repairs are performed. Personnel should be familiar with the processes involved in starting up a system and shutting in a pipeline segment. The general steps involved in each process are covered in this section.

Personnel who perform a system startup or shutdown must possess basic knowledge in the following areas:

- Pipeline pneumatics
- The MAOP for the pipeline system
- Company operations policies
- Company policies for the appropriate response to problems

In addition, personnel must be familiar with the pipeline's normal operations. They must know what is normal so that they can quickly spot anything that is not normal.

8.1.0 Startup

There must be a clear flow path from the origin point to the delivery point before the pipeline is started up. All equipment must be functioning properly and ready to receive the gas flow, which moves downstream from the origin point. Compressors at downstream stations are activated by a pressure rise from upstream stations.

Specific steps for a startup procedure and the order in which the steps are taken vary from one company to another. In general, a typical startup may include some or all of the following steps:

Step 1 Confirm that the proper work orders and work schedules have been issued.

Step 2 Confirm that the gas is ready to enter the pipeline.

Step 3 Confirm that upstream and downstream stations are ready for startup. This step includes confirming that all necessary equipment is available.

Step 4 Confirm that all main-line station valves are in the normal position. Confirm that the flow path is correct.

Step 5 Start the compressors.

Step 6 Notify the next station downstream that the compressors are running.

Step 7 Start the compressor units at stations on the line that will be affected immediately.

Step 8 Notify the next station downstream that the compressor unit at that station is coming online.

Step 9 Set the compressor unit to the desired pressure. Watch for the rise in pressures downstream.

Step 10 Notify each station downstream that the compressor unit is coming online.

Step 11 Monitor the system until the pipeline is operating in a **steady state.**

Step 12 Ensure that all required documents have been completed.

8.2.0 Shut In

Sometimes a pipeline section must be shut in, or isolated, for inspection, maintenance, or repair work. This work can be performed without stopping the flow of gas by installing a **hot tap.** Hot tapping requires a **hot work** permit, and personnel are not allowed to perform this task until all the correct permits are obtained. The hot tap process allows personnel to install a loop of pipeline so that the gas can be diverted around the shut-in section. This loop is welded to the main pipeline with valve fittings at either end.

When a hot tap is installed, another fitting is welded on the main pipeline near the section to be shut in. A valve and tapping device are installed on this fitting, and the valve is opened. A pilot drill attached to the tapping device drills a hole in the pipeline. Gas moves from the pipeline into the tapping device, air is **purged,** and the valve is closed. Any remaining gas in the shut-in section is vented to the atmosphere. The valves on the loop pipeline are opened, allowing the gas to take a detour around the shut-in section. The shut-in section is checked to ensure that all gas and air have been purged before any repairs or inspections are performed. Depending on the circumstances, the loop may be a permanent or a temporary installation. If the loop is temporary, the process is reversed to shut in the loop and return flow to the main line.

Specific steps for shutting in a pipeline segment and the order in which the steps are taken vary from one company to another. Additional steps are required for a full system shutdown (if one should ever be required), and each company has its own procedures for this process. In general, a typical shut-in procedure may include the following steps:

Step 1 Confirm that the proper work orders, work permits, and work schedules have been issued.

Step 2 Complete maintenance on valves if necessary.

Step 3 Conduct a leakage survey over the segment to be shut in.

Step 4 Confirm that affected personnel are prepared to shut in the pipeline segment.

Step 5 Install a hot tap and isolate the section to be shut in.

Step 6 Install the loop and monitor pressure as gas flows into the loop.

Step 7 Notify appropriate personnel that the pipeline segment is shut in.

Step 8 Monitor pressures in sections adjacent to the shut-in section to ensure the MAOP is not exceeded.

Step 9 Complete the required maintenance, repair, or inspection.

Step 10 Reverse the process to remove the loop and restore gas flow to the isolated section if appropriate.

Step 11 Submit the required reports.

It is important to monitor the area for leaks during the entire shut-in process. Personnel must wear the appropriate PPE while performing this task.

8.3.0 Monitoring

Conditions on a pipeline can change without warning. That is why it is important to monitor conditions when a segment is shut in. Field personnel rely on the control center to monitor pipeline pressure and temperature to ensure they remain within acceptable limits. They also rely on safety devices at the work site to alert them to problems.

If there is an unexplained change in pressure, the general procedure is to determine the cause first and then control the situation. The control center may make adjustments to bring the pressure back to acceptable limits or alert field personnel to problems they must address. Field personnel may be directed to check any compressors or valves operated before the pressure change to ensure they are still in the right position. Similar procedures are followed in response to any other problems that could affect the safety of personnel or the integrity of the pipeline.

9.0.0 ◆ PURGING

Sometimes a pipeline section must be isolated so that it may be purged for inspection or maintenance. Gas is routed around the isolated section so that there is no interruption in service. When isolating a pipeline section, personnel must **unpack** and purge the contents. After the pipeline is emptied of gas, explosive vapors may still be present. Personnel must then purge the line to remove all hydrocarbon gases. Then personnel must do a second purge. They must purge the air in the pipeline with natural gas before repressurizing and **packing** (*49 CFR 192.69*).

Flammable gas explodes or catches fire when it is mixed with a certain percentage of air. Typically, a mixture of gas and air must be 5% to 15% gas and 85% to 95% air before it explodes. A mixture of gas and air that does not have enough gas to explode is below the LEL. It is too lean to explode.

A mixture with more than the maximum amount of gas is above the **upper explosive limit (UEL).** It is too rich to explode. Purging establishes a moderately fast flow in the line being purged. This rapid flow helps to ensure that where the air and gas meet, they mix as little as possible. Personnel must maintain a constant purge rate until all air or gas has been expelled.

9.1.0 Purging Gas from a Pipeline

To purge a transmission pipeline segment, personnel open valves and use an inert gas to push out the flammable gas. The employee in charge of the purging procedure defines the appropriate process to be followed. All employees engaged in the process must wear the appropriate PPE. Before and during purging operations these personnel must do the following tasks:

- Follow the company operations and maintenance procedures and be aware of possible hazards along the pipeline ROW.
- Monitor for hazardous vapors throughout the purging process.
- Know and follow the company's lockout/tagout procedure.
- Know and follow the company's procedures for obtaining the appropriate work permits.

Following are the basic steps involved in purging gas from a pipeline:

Step 1 Safely shut in, vent, and disconnect the pipeline section that must be purged.

Step 2 Unpack and purge the segment.

Step 3 Purge the air in the pipeline with natural gas before repressurizing and packing the line.

Step 4 Reconnect all equipment after the purging is completed.

Step 5 Submit the required reports.

Local distribution companies may have additional procedures to purge the distribution lines that serve homes and businesses. Their procedures also include a process for quickly and safely restoring service.

9.2.0 Isolating the Section to Be Purged

Isolating the section to be purged requires following the company's lockout/tagout procedure. This procedure must be followed when hazardous materials could adversely affect maintenance. It includes isolating all such materials from the work area. The isolation must continue until the work is completed. Personnel must close off all valves and other devices that are directly connected to the section being isolated. These devices must not be accidentally opened while the purging operation is occurring.

Bonding provides a safe path for electric current across any gap in the pipeline while the purging is occurring. Sparking or arcing should be prevented during the purging process.

Personnel must measure the explosive level of hazardous vapors with a CGI. No one should rely on sense of smell to detect hazardous vapors for the following reasons:

- The concentration levels may not be high enough to detect dangerous levels by the sense of smell alone.
- A medical condition may affect a person's sense of smell.
- Certain hazardous vapors may deaden the sense of smell.

9.3.0 Purging Air from Pipelines

When there is air in a pipeline that could form a hazardous mixture with natural gas, that air must be purged from the pipeline. This process involves opening valves and using natural gas to push out the air. The procedures for purging air from a pipeline are the same as those for purging gas with the exception that you must sample the vent stream with a combustible gas indicator or portable gravitometer to determine when the vapor reaches 100% gas in air. Personnel must follow all of the same safety procedures.

9.4.0 Monitoring During Purging

During the purging process, personnel must monitor a number of conditions to ensure that the process is carried out safely. The following conditions are monitored:

- The air for hazardous gases
- The weather, especially wind direction
- All communications from the control center, especially those that alert personnel to hazardous conditions

9.5.0 Abnormal Operating Conditions

Several AOCs are associated with the purging operation. Following are examples of common AOCs:

- *Gas leak*—A hazardous gas alarm will sound. Personnel may be required to leave the area. The gas leak must be repaired and the air checked with a CGI to ensure safety.

- *Hazardous vapors*—Wind may carry hazardous vapors, so pipeline personnel must take special care when conducting a purging operation in a populated area. They must monitor vapors to ensure that they are not in the explosive range. Purging operations must be suspended if weather conditions do not allow the vapors to safely disperse.
- *Fire or explosion*—Personnel must carry out company safety procedures to protect people, property, and the environment. The area may have to be evacuated. Authorized company personnel may contact local emergency response agencies if necessary.

Review Questions

Sections 7.0.0–9.0.0

1. When a safety device alarm goes off, it must _____.

a. be reset immediately
b. never be ignored
c. be turned off immediately
d. be monitored only in the control center

2. By law, the flammable gas detection devices must signal an alarm when gas levels reach _____ percent of the LEL.

a. 10
b. 15
c. 20
d. 25

3. The _____ process allows personnel to install a loop of pipeline so that the gas can be diverted around the shut-in section.

a. hot tap
b. pigging
c. tie in
d. detour

4. A mixture of gas and air that does not have enough gas to explode is too _____.

a. rich
b. odorized
c. compressed
d. lean

5. _____ provides a safe path for electric current across any gap in the pipeline while purging is occurring.

a. Hot tapping
b. Cold tapping
c. Bonding
d. Splicing

10.0.0 ◆ PIGGING

A **pig** is a device that is inserted into a pipeline to perform a variety of functions that cannot be performed by personnel. Most pigs are maintenance devices, but pigs serve many functions other than cleaning the inside of a pipeline. They also push out water or air during testing, remove liquid from gas pipelines, and monitor a pipeline's interior condition for the buildup of substances or corrosion.

The size of the pig depends on the size of the pipeline for which it is designed. Some pigs are as much as 42 inches in diameter. Lengths vary from a few inches to several feet (see *Figure 16*). Pigs are always larger in diameter than the pipeline they are to run in, so that the cups on their surfaces are compressed against the pipeline's walls. This compression aids in the pig's ability to scrape material from the pipeline walls.

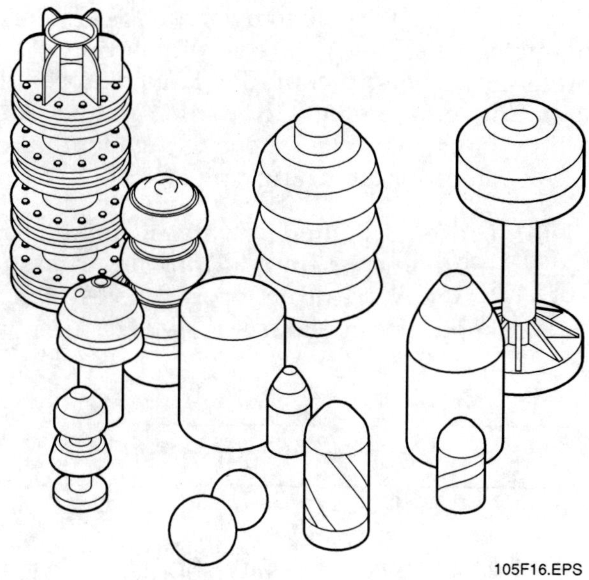

105F16.EPS

Figure 16 ◆ Assortment of pigs.

Pigs aid in the removal of debris, such as rust, scale, or wax. Buildup of such debris encourages corrosion and restricts the efficiency of the pipeline. Each pig is selected for the specific job that is needed in the pipeline. Using pigs keeps corrosive damage to a minimum and keeps pipeline flow as free as possible. If a pipeline is restricted, sending a pig through the line usually restores full or near-full flow. Pipelines that transport clean products, such as natural gas, usually require fewer pigging operations than pipelines that transport more contaminated products, such as crude oil.

Smart pigs can perform all kinds of electronic surveillance, including an exact measurement of the pipeline's inner surfaces. They can check the thickness of pipeline walls and find interior damage. Electronic signals transmitted to operations personnel make it easy to track, direct, and retrieve smart pigs.

10.1.0 Launching, Tracking, and Receiving Pigs

Pigging operations are scheduled as regular maintenance on most pipelines. The pigs travel with the product flow and are pushed along by the pipeline pressure. They travel much like a piston in an endless hydraulic cylinder. As they travel, they scrape, brush, and clean.

10.1.1 Launching the Pig

The methods for **launching** a **pig run** are the same for any type of pig, but they vary with the size of pig. Small pigs are loaded by hand. Pigs for large-diameter pipelines require several employees or, preferably, a hoist or crane. Pig launch procedures vary from one company to another, and you must follow your company's policy. In general, these are the steps for launching a pig:

Step 1 Close all launching valves, isolating the pig launcher from all pipeline pressure. Valve 3 should be open. Valves 1 and 2 should be closed (see *Figure 17*).

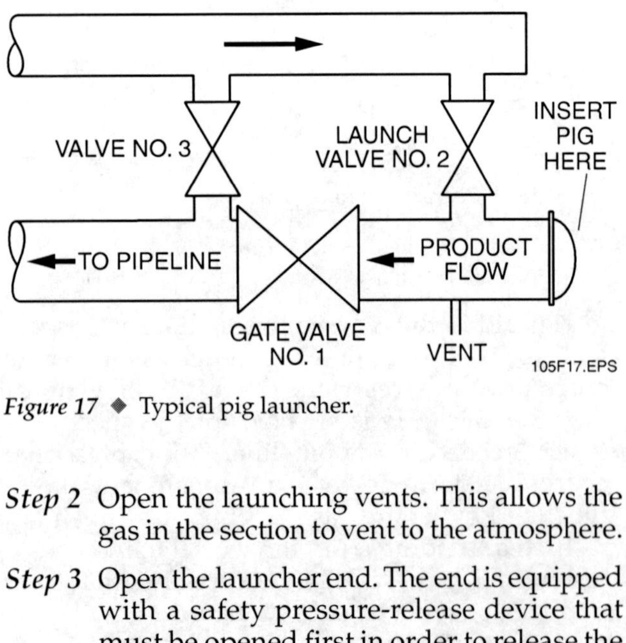

Figure 17 ◆ Typical pig launcher.

Step 2 Open the launching vents. This allows the gas in the section to vent to the atmosphere.

Step 3 Open the launcher end. The end is equipped with a safety pressure-release device that must be opened first in order to release the end cover. The release devices work differently depending on the manufacturer.

Step 4 Load the pig into the barrel.

Step 5 Repressurize the launcher by opening the launching valves until the pressure is equalized. At this point, the pig is ready to be launched.

Step 6 When the signal to launch the pig is received, simultaneously open valves 1 and 2 and close valve 3 to launch the pig.

Step 7 Open valve 3, then close valves 1 and 2 after the pig has moved into the main flow of the pipeline. This step isolates the pig launcher again.

Step 8 Record all procedures on the appropriate report form.

10.1.2 Tracking the Pig

Once the pig has been launched, operations personnel downstream must be told about the type of pig, when it was launched, and roughly when it will arrive. Most stations have some kind of remote device to signal that the pig has arrived.

Pig tracking is vital to the operation of the pipeline because trouble can occur quickly. Pigs can get stuck or hung up in the pipeline, or become so damaged that they cannot travel or can travel only at a reduced speed. Such pigs must be found immediately to prevent serious problems in the pipeline.

If a problem develops, operations personnel may notice a sudden change in pressure. The pig may be late or fail to arrive. When personnel are certain that the pig has not arrived, they must notify the control center immediately. Some pigs are equipped with electronic signaling devices that help pinpoint their location. For pigs without signaling devices, a search must be initiated.

A typical search begins by checking all station piping and equipment. If the pig is hung in a valve, personnel may hear a noise from inside the pipeline or valve. If the pig has been damaged or has self-destructed, pieces of the pig may be found in the strainer.

If the pig is not in the station piping, the search must move upstream. The first place to look is the launcher at the original station to ensure that the pig left the station. If the pig is not found upstream, the search must move to the downstream section of the pipeline.

Once the pig is located, a second pig may be launched in an effort to bump or push the first pig loose. The danger is that the second pig may also become stuck. The second pig should be radio equipped so that its progress may be followed closely. If the second pig is able to push the first one loose, the two must be monitored closely to ensure that both pass through the pipeline.

10.1.3 Receiving and Unloading the Pig

Once the pig has been launched, information about its type and schedule must be sent to operations personnel downstream. A detector signal should be sent to the control center of the next station, warning that the pig is approaching. The valve alignment must be prepared to receive the pig (see *Figure 18*).

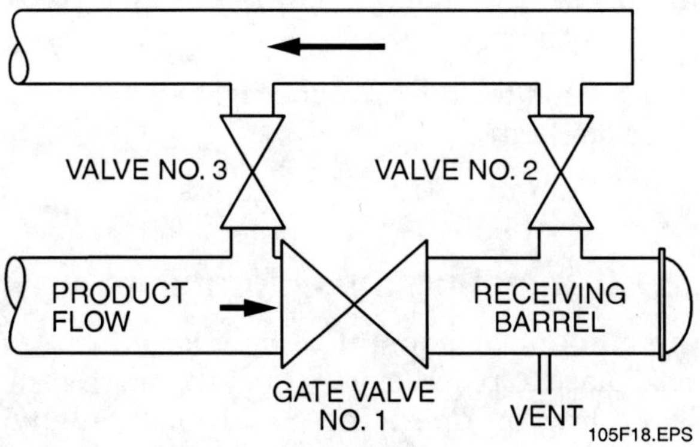

Figure 18 ◆ Pig-receiving installation.

The procedure for receiving pigs may vary from one company to another, and you must always follow your company's procedures. In general, these are the steps that must be followed when receiving a pig:

Step 1 Set up the receive valve alignment as soon as possible by opening valves 1 and 2 and closing valve 3 (refer to *Figure 18*).

Step 2 Listen for the pig to arrive at the receiving unit.

Step 3 Ensure that the restraining bar is in place to keep the pig from going back into the pipeline.

Step 4 Open valve 3 and close valves 1 and 2.

Step 5 Vent gas to the atmosphere.

Step 6 Release all pressure before attempting to open the receiver end cap.

Step 7 Pull the pig from the receiving barrel.

Step 8 Place the pig on a concrete slab for inspection.

Step 9 Inspect the pig.

Step 10 Complete required reports and send them to the control center. The reports should document the pig's condition and all test results.

11.0.0 ◆ UPRATING THE PIPELINE MAOP

Uprating is the term used to describe the process for increasing the MAOP on an existing pipeline segment. The MAOP is the highest pressure allowed for a pipeline segment. Various factors, such as pipeline design and operating experience, are considered when setting the MAOP. Engineers who have been trained on pipeline design and safe operating limits handle uprating the MAOP. However, it is useful for all pipeliners to have a basic understanding of this process.

Step-by-step pressure increases are necessary to safely uprate a pipeline. A trained individual making such pressure increases must know a great deal about pipeline operations. This knowledge includes the following areas:

- A thorough knowledge of the uprating plan
- A basic knowledge of the system being uprated
- An understanding of company safety policies and procedures

11.1.0 Performing Step-by-Step Pressure Increases

Before performing step-by-step pressure increases, it is essential to increase the pressure in the pipeline segment to the current system MAOP. Doing so provides a starting point for the step-by-step pressure increases. Then the pipeline pressure must be increased gradually to the first planned setting.

When the pipeline pressure reaches the first planned setting, the process must halt before the pressure is increased to the next setting. Gradual, step-by-step pressure increases help to prevent pipeline failure and accidents. Each pressure increase must be recorded on a pressure test form.

The pipeline must be monitored constantly as the pressure increases. A pressure loss could indicate a leak. If there is a pressure loss, a leak survey must be conducted to check for possible leaks. Leaks must be monitored or repaired as specified in *49 CFR Subpart K*.

Step-by-step increases in pipeline pressure should be continued, stopping at each setting as planned, until the final planned test pressure is reached. When the final pressure is reached, the pipeline is uprated.

11.2.0 Documenting the MAOP

The final established MAOP must be documented. The following information must be included:

- The name of the operating company
- The name of the contractor
- The name of the person conducting the uprating test
- The date and time of the test
- The location of the pipeline or the pipeline segment uprated
- The test medium used
- The test pressures
- The duration of the test
- The uprating plan
- Any leakage and repairs
- Any AOCs and the corrective actions taken

11.3.0 Abnormal Operating Condition

If a leak occurs during an uprating, the test must be stopped and a determination made whether to repair the leak or continue the test while monitoring the leak. Each detected leak must be repaired or monitored in accordance with Department of Transportation regulations (*49 CFR Subpart K*). Personnel must carry out company safety procedures to ensure the safety of those who are performing the uprating procedure.

12.0.0 ◆ OPERATING ODORIZATION EQUIPMENT

Natural gas is odorless. For safety reasons, a federal regulation requires that a person with a normal sense of smell must be able to easily detect natural gas when the gas concentration is only 20 percent of its LEL (*49 CFR 192.625*). Therefore, an **odorant,** usually a chemical called mercaptan, is added to the natural gas stream in distribution lines. Generally, local distribution companies are responsible for adding the odorant, which gives the gas a distinctive rotten-egg smell.

Personnel responsible for injecting odorant into a natural-gas stream should read and follow the material safety data sheet for the odorant being used. In addition, employees must follow established company procedures for injecting odorant and wearing the appropriate PPE.

The steps for injecting odorant may vary from one company to another. Generally, these are the basic steps for operating odorization equipment:

Step 1 Ensure that the odorant is the proper one for the intended use.

Step 2 Operate the odorization equipment so that the odorant is injected smoothly and consistently.

Step 3 Measure the odorant level in the pipeline system.

Step 4 Monitor the odorant level in the pipeline system.

Step 5 Submit the required reports.

12.1.0 Injecting the Odorant

The amount of odorant to be injected can vary from place to place, even within the same system, depending on operating conditions. Following are three types of equipment that may be used to inject the odorant:

- *Bypass type*—Diverts the main gas stream through a mechanical device to disperse the liquid odorant into the gas
- *Wick type*—Uses the flow of gas across a wick or **pitot tube** to pick up liquid odorant for dispersion in the gas stream
- *Electronic*—Uses an onboard microprocessor to provide a flow measurement signal and generate data that is stored in an audit trail

12.2.0 Monitoring the Odorant

Once the odorant has been injected into the gas, personnel must monitor it to ensure that proper ratios are maintained. The type of equipment used for this purpose depends on company policy. Generally, personnel use an **odorometer** (see *Figure 19*). This meter allows personnel to determine if the ratio of the usage to the volume of gas is within acceptable limits. Odorometers must be properly calibrated to produce valid measurements. Personnel must follow the manufacturers' recommendations for proper use.

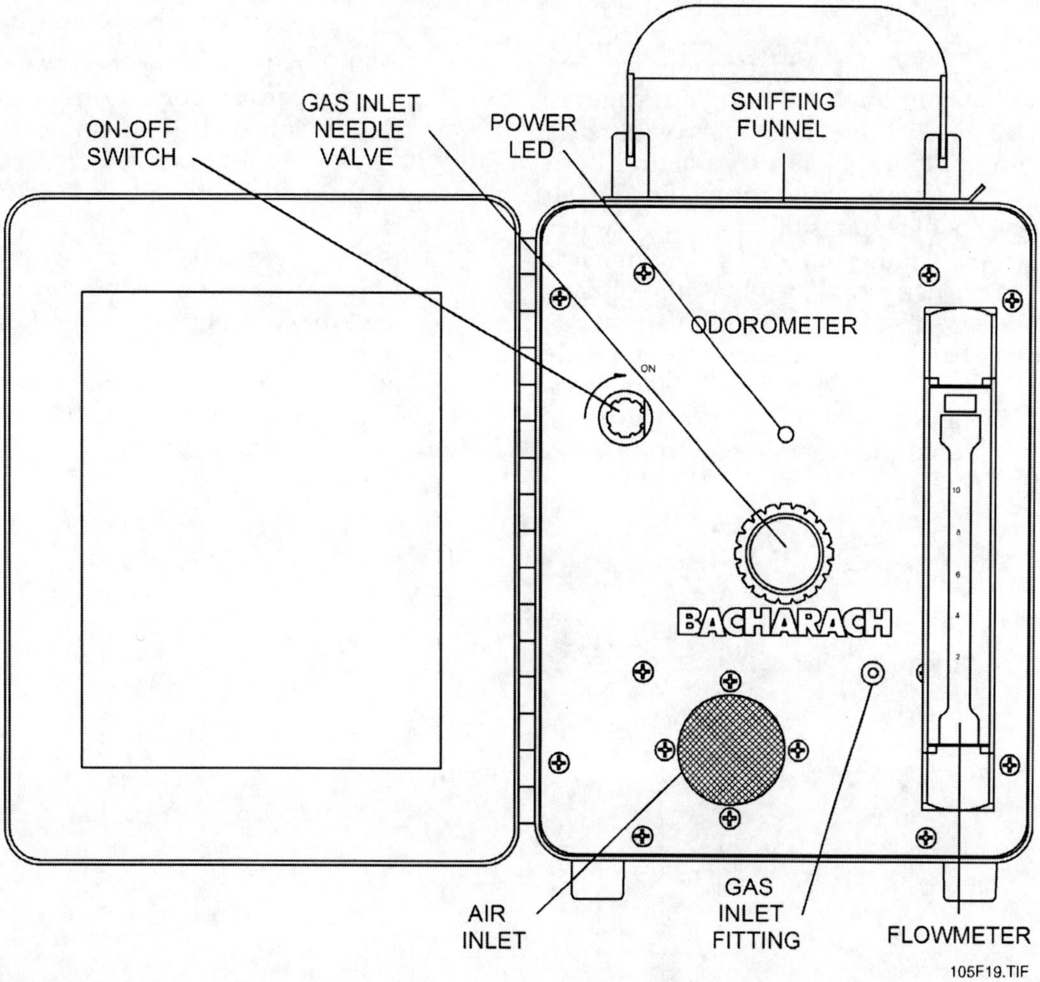

Figure 19 ◆ Odorometer.

Review Questions

Sections 10.0.0–12.0.0

1. Other than cleaning the inside of a pipeline, pigs can also _____.
   a. perform minor repairs
   b. monitor the interior condition of the pipeline
   c. odorize the gas in the pipeline
   d. uprate the pipeline

2. Pig _____ is vital to the operation of the pipeline because pigs can get stuck or become damaged.
   a. launching
   b. receiving
   c. loading
   d. tracking

3. Engineers who have been trained on pipeline design and safe operating limits handle _____.
   a. pigging operations
   b. purging pipeline segments
   c. uprating the MAOP
   d. monitoring odorant in distribution lines

4. When _____ a pipeline, an essential step is to increase the pressure to the current MAOP.
   a. purging
   b. uprating
   c. pigging
   d. cleaning

5. Natural gas is usually odorized by _____.
   a. the local gas distributor
   b. operations personnel
   c. the pipeline company
   d. a contractor

Summary

In this module, you learned that routine control center operations are often anything but routine. Working in the control center involves overseeing many pipeline operations and interacting with personnel from all areas of the company. The job also involves a lot of responsibility. Control center work can be challenging but also satisfying. You are part of an industry that provides a necessary and welcome service to its customers. Natural gas is an energy-efficient and clean product that makes our modern lifestyle possible. Many people rely on it for warmth and comfort in their homes, and many industries rely on it for a wide variety of processes.

Although natural gas is generally regarded as safe, under certain conditions it can be a hazardous substance. Transporting it in pipelines involves some risks for pipeline workers, the public, and the environment, which is why you must always do your job safely and responsibly.

This module provided you with basic knowledge of the tasks you must perform and the skills you need for control center operations. Although you may not perform all of the tasks outlined in this module, you must have a basic understanding of them so that you can carry out your duties in the control center effectively.

GLOSSARY

Trade Terms Introduced in This Module

Abnormal operating condition (AOC): A condition within a pipeline system that exceeds operating design limits or that may result in a hazard to persons, property, or the environment (*49 CFR 192.803*).

Blowdown: The depressurization of gas piping, which is accomplished by releasing the pressure at a safe atmospheric vent location.

Bonding: The permanent joining of metallic parts to form an electrically conductive path, which ensures a safe electrical connection.

British thermal unit (Btu): The amount of heat required to raise one pound of water one degree Fahrenheit.

Combustible gas indicator (CGI): An instrument used to sample the atmosphere to find out if it is explosive.

Compressor station: Facilities installed along pipelines approximately every 100 miles that compress natural gas to reduce its volume and force it through the pipeline.

Control center: The operations center of a pipeline, usually located in a terminal or main pipeline facility. All the control consoles and computer-operated systems are located in the control center.

Derate: To reduce the rating of a pipeline or other device to permit operation at parameters other than those originally established.

Fire suppression material: A chemical or gas used to smother fires quickly.

Hot tap: A process in which a drill or circular cutter is used to cut a hole in a pipeline in which gas is still flowing to shut in, or isolate, that pipeline segment.

Hot work: Any activity that could produce flames, sparks, or other hot fragments that might act as an ignition source to flammable materials in the area. The activities covered by this term include but are not limited to welding, flame cutting, and soldering.

Ionization detector: A photoelectric detection device that signals an alarm when it detects smoke.

Launching: Releasing a pig into a pipeline; it is also known as kicking.

Linewalker: A member of the ground patrol that inspects a pipeline's right-of-way.

Lower explosive limit (LEL): The minimum concentration of vapor in air below which fire does not occur in the presence of an ignition source.

Manifold: A pipeline fitting with several side outlets to connect it with other piping. It is also known as a header.

Odorant: A liquid added to natural gas to give it an odor. Because natural gas is odorless, odorant is added so that the gas can be detected by smell.

Odorometer: An instrument that is used to measure the odorant level in a gas.

Pack: To compress natural gas under pressure in a pipeline.

Pig: A scraping tool with blades or brushes that is forced through a pipeline to clean out accumulations of wax, scale, and debris from the pipeline walls. It is also called a line scraper or a go-devil.

Pig run: A pig's trip through a pipeline.

Pig tracking: The process of monitoring the progress of a *pig* as it travels through the pipeline.

Pipeline segment: A specified length of pipeline constructed using a specified grade of piping.

Piping and instrumentation diagrams (P&IDs): Engineered diagrams of piping, valves, and electrical utilities.

Pitot tube: A tube placed in a gas or fluid flow to measure velocity.

Programmable logic controller (PLC): A small industrial computer that monitors individual circuits that relay information about pipeline parameters and sends an alarm when parameters are exceeded.

Purge: To replace the atmosphere in a pipeline by injecting an inert substance to prevent formation of explosive mixtures.

Remote terminal unit (RTU): A computerized unit that transmits the status of valves to the control center and receives commands from the control center to the valves.

Remotely controlled shutdown device (RCSD): An electronic device that is used to shut down a compressor driver from the control center.

Self-contained breathing apparatus: A device consisting of a face piece and an air canister mounted in a shoulder harness that enables workers to enter areas where there is not enough oxygen or where there are dangerous concentrations of fumes or gases.

Smart pig: An electronic device inserted into a pipeline that senses and records irregularities in the line. It is sometimes called an instrumented pig.

Steady state: The point at which pressures and flows are relatively constant over time.

Thermal detector: A heat-sensing device that warns of the presence of flame by detecting a sudden rise in temperature. It is also known as a flame detector or a bolometer.

Ultraviolet (UV) detector: A sensor that notes the presence of fire by detecting the ultraviolet rays given off by fires that are just beginning.

Unpack: To decompress gas that is under pressure in a pipeline.

Upper explosive limit (UEL): The maximum concentration of vapors in the air that will burn when an ignition source is present.

Uprating: Increasing the maximum allowable operating pressure on an existing pipeline segment.

Valve alignment: The exact sequence of valves that must be opened and closed at the same time so that gas can be received at and delivered to specific locations.

Valve interlocking: The process of operating a series of valves in a particular sequence. For example, certain valves will not close until another key valve has closed. A computerized hybrid-logic program controls the process.

Voted system: An alarm system layout used for flame detection and gas detection, employing sensitive detectors and delayed action relays. Such systems will not alarm until two or more detectors are tripped.

Figure Credits

Gardner Denver	105F10
Stewart and Stevenson	105F11
Matheson Tri-Gas, Inc.	105F12
Barcharach, Inc.	105F19

CONTREN™ LEARNING SERIES — USER UPDATES

The NCCER makes every effort to keep these textbooks up-to-date and free of technical errors. We appreciate your help in this process. If you have an idea for improving this textbook, or if you find an error, a typographical mistake, or an inaccuracy in NCCER's Contren™ textbooks, please write us, using this form or a photocopy. Be sure to include the exact module number, page number, a detailed description, and the correction, if applicable. Your input will be brought to the attention of the Technical Review Committee. Thank you for your assistance.

Instructors – If you found that additional materials were necessary in order to teach this module effectively, please let us know so that we may include them in the Equipment/Materials list in the Instructor's Guide.

Write: Curriculum Revision and Development Department
National Center for Construction Education and Research
P.O. Box 141104, Gainesville, FL 32614-1104

Fax: 352-334-0932

E-mail: curriculum@nccer.org

Craft ____________________ Module Name ____________________

Copyright Date __________ Module Number __________ Page Number(s) __________

Description

____

____

____

____

(Optional) Correction

____

____

____

(Optional) Your Name and Address

____

____

____

Module 67106-02

Quality Control/ Measurement

Introduction to the Pipeline Operator Qualification Rule, API, and NCCER

It could be said that petroleum products keep America running. Petroleum products like oil and natural gas are the principal sources of energy in the United States and are used in many manufacturing and chemical processes. Petroleum products are primarily transported by pipeline. There are more than a million miles of liquid and natural gas pipeline in this country, ranging in diameter from two inches to four feet. These pipelines bring crude oil and natural gas from production fields to processing plants and refineries—and ultimately to the consumer.

Pipeline Operator Qualification

Petroleum products are considered hazardous, so pipeline personnel must take extra precautions and follow specific procedures when working with them. To help ensure the safe handling and transportation of petroleum products, the United States Department of Transportation (DOT) mandates safe practices as federal regulations.

One such regulation, the Operator Qualification (OQ) rule, requires pipeline operators to develop a written qualification plan to evaluate their personnel and their contract personnel (see *49 CFR Part 192 Subpart N—Gas*, and *49 CFR Part 195 Subpart G—Liquid*). The evaluation hinges on the operator's ability to perform certain covered tasks and to be able to recognize and respond to abnormal operating conditions that may be encountered while performing these covered tasks.

Covered Tasks

A pipeline performance task is deemed *covered* if the activity passes the following four-part test:

1. Is it performed on a pipeline facility?
2. Is it an operations or maintenance task?
3. Is it performed as a requirement of *49 CFR Part 195 (Liquid)* or *49 CFR Part 192 (Gas)*?
4. Does it affect the operation or integrity of the pipeline facility?

If the answer to all four questions is *yes*, then the activity is considered *covered* and must figure into the evaluation process of pipeline personnel.

American Petroleum Institute

The OQ rule and its related covered tasks provided the catalyst to develop a unified pipeline training and assessment program—so that pipeline operators and owners have ready access to the assessment and training tools they need. This effort was the direct result of the vision and leadership provided by The American Petroleum Institute (API) and API members.

Efforts to unify the industry are nothing new to API. Indeed, the impetus for forming the Institute in 1919 was the need to standardize engineering specifications for drilling and production equipment. Today, the most pressing issues revolve around public perceptions and government policies toward the industry. Speaking with one voice on these issues has become as essential as having interchangeable parts in the field. API's members determine the petroleum industry's positions on public policy as well as the standards that govern its day-to-day operations worldwide.

Consortium on Operator Qualification

API assembled a focus group whose primary purpose was to identify covered tasks and provide compliance recommendations to pipeline operators.

The Consortium on Operator Qualification (COOQ) became the guiding force for the identification and development of covered tasks for the liquid pipeline industry. Their work product, API publication 1161, *Guidance Document for the Qualification of Liquid Pipeline Personnel*, formed the basis for the resulting Pipeliner Training and Assessment Program.

The covered task list created in API publication 1161 has now evolved to include gas covered tasks and is maintained by NCCER. The most current covered task list can be downloaded at www.nccer.org.

National Center for Construction Education and Research

With the guidance and support of API, its members, and the Consortium on Operator Qualification, the National Center for Construction Education and Research (NCCER) was able to design and implement the Pipeliner Training and Assessment Program.

NCCER is a not-for-profit 501(c)(3) education foundation established in 1995 by the world's largest and most progressive construction and maintenance contractors and national contractor associations. NCCER was founded to address the

severe workforce shortage facing the construction and maintenance industries and to develop a standardized training process and curricula. Today, NCCER is supported by hundreds of leading construction and maintenance companies, manufacturers, and national associations and developed more than 35 construction and maintenance titles.

NCCER also maintains a National Registry, which contains the records of trainees and instructors who have received NCCER training. All completion information for every trainee is sent to the NCCER and filed within the Registry. The Registry can then confirm training and skills for workers as they move from company to company, state to state, or even within their own company.

Standardized Curriculum (NCCER's Contren™ Learning Series)

NCCER curricula are developed using the expertise of industry subject matter experts (SMEs). The use of SMEs as an integral part of every new development or revision ensures that the curriculum is current, industry-driven, and indicative of industry practices. The development and/or revision process is facilitated from start to finish by NCCER project managers, who work closely with the SMEs and a select team of technical writer(s) on the development and presentation of the curriculum.

Pipeline: Five Training Paths

The pipeline curriculum focuses on essential skills and competencies within five skill areas:

- Corrosion Control
- Electrical and Instrumentation
- Maintenance
- Mechanical
- Operations (Field and Control Center)

SMEs from the pipeline industry have determined the skills and competencies needed based on API's covered tasks and the consensus of the group/industry. Specific company information is generally not incorporated into the curriculum unless the SMEs determine that the information is the industry standard.

Training Modules and Covered Tasks

The smallest unit of instruction in the curriculum is a *module*. Each curriculum is composed of modules (the equivalent of stand-alone chapters) that are logically sequenced into complete *levels* of instruction. Every module is task-oriented and requires the successful completion of a written (knowledge-based) examination and a performance (hands-on) test. Modules may be individually selected and/or combined for efficient task training. Modules in the pipeline curricula, to the extent possible, will align with API-developed covered tasks. While not all modules will relate to covered tasks, all covered tasks will relate to at least one module.

Benefits of Training to the Pipeline Marketplace

Applying NCCER's standardized training process allows pipeline owners and contractors in the United States to establish the benchmark for acceptable skill and competency levels within the pipeline industry. This process will assist training organizations in developing and/or improving standardized pipeline training programs that effectively match their resources and operating conditions while responding to regulatory demands and to the needs of the industry. The process will also recognize training organizations that meet the minimum standards of competence, warranting regulatory, public, and professional confidence. This in turn will foster national unity within the pipeline industry for the mutual benefit of individual technicians, owners, contractors, and ultimately pipeline-users and society.

Additional Resources

To review the Pipeliner Training and Assessment Program Implementation and Resource Guide and the latest version of pipeline covered tasks, visit NCCER's Web site at www.nccer.org, or contact:

National Center for Construction Education and Research
PO Box 141104
Gainesville, FL 32614-1104
(888) 622-3720
Web site: www.nccer.org

For more information on API, please contact:

American Petroleum Institute
1220 L Street, Northwest
Washington, D.C. 20005
(202) 682-8000
Web site: www.api.org

To order *API Pub 1161*, contact Global Engineering Documents by calling 1-800-854-7149 or visiting their Web site at www.global.his.com.

Additional sources of information on the pipeline operator qualification regulation include:

DOT – Office of Pipeline Safety
www.ops.dot.gov

Transportation Safety Institute
www.tsi.dot.gov

COURSE MAP

This course map shows all of the modules in the Gas Pipeline Operations curriculum. The suggested training order begins at the bottom and proceeds up. Skill levels increase as you advance on the course map. The local Training Program Sponsor may adjust the training order.

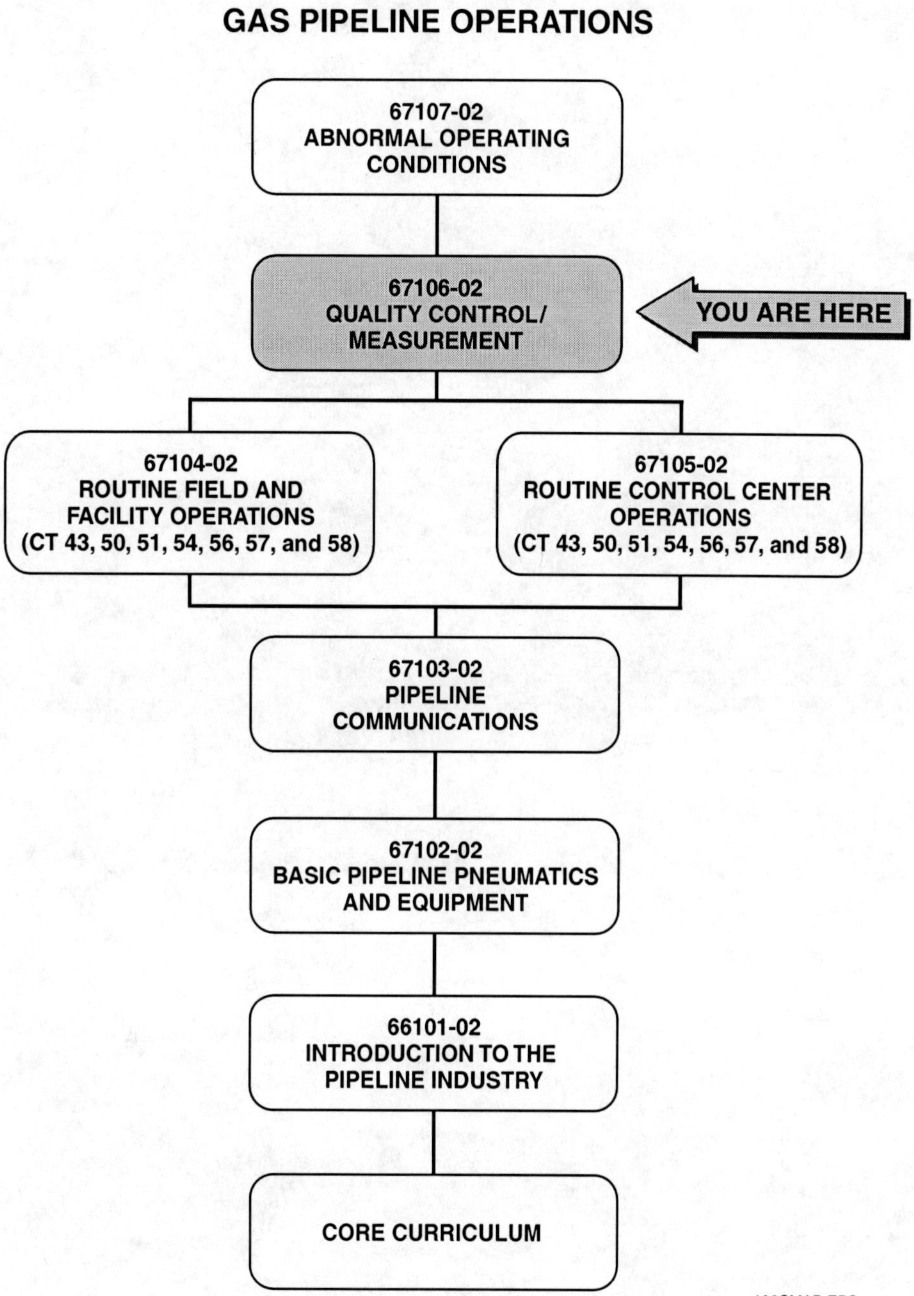

106CMAP.EPS

MODULE 67106-02 CONTENTS

Figures

Tables

MODULE 67106-02

Quality Control/ Measurement

Objectives

When you have completed this module, you will be able to do the following:

1. Take product samples.
2. Perform product testing.
3. Identify types of meters.
4. Define measurement and understand how and why gas is measured and how to interpret measurements.
5. Verify the accuracy of meters and test gauges.
6. Explain the reasons for injecting and monitoring odorant.

Prerequisites

Before you begin this module, it is recommended that you successfully complete the following: Core Curriculum; Pipeline Core, Module 66101-02; Gas Pipeline Operations, Modules 67102-02 and 67103-02 and 67104-02 or 67105-02.

Required Trainee Materials

1. Pencil and paper
2. Appropriate personal protective equipment

1.0.0 ◆ INTRODUCTION

Quality control and field measurement are important for ensuring the safe, reliable delivery of natural gas. Careful work in these areas is also important for the financial health of the pipeline industry. This module introduces you to these operations.

Government and industry standards apply to all work done in the pipeline industry. Quality control employees do many jobs to help ensure that the gas meets the highest possible standards. These employees apply quality standards, ensure that government and industry standards are met, and solve problems when they occur.

Quality control personnel take samples and perform tests for the following reasons:

- To check quality
- To remove unwanted elements from the gas
- To monitor odorant levels where appropriate

Odorant is added to natural gas distribution lines. Usually, the local gas distributor is responsible for adding and monitoring odorant. Government regulations state when and where odorant must be added. Where gas pipelines pass through thinly populated areas, for example, regulations may not require an odorant. In addition, odorant generally is not required in gas transmission pipelines.

Accurate measurement is at the heart of the pipeline industry's accounting system. Companies bill their customers based on how much gas they order. Customers naturally want to pay only for what they actually receive. It may seem simple enough to receive an order, send the right amount of gas, make a record of the transaction, and bill the customer. However, two types of losses can affect accurate measurement: actual loss from a leak, accident, or theft and what is known as book loss, which results from inaccurate measurement, recording mistakes, or other clerical errors.

In addition, measuring gas is not a simple process. Gas volume changes with changes in temperature and pressure. There is no single measuring process that works on all forms of natural gas. For example, gas under pressure in an underground well is measured using a process different from that used for gas in a high-pressure pipeline. Different processes are needed because the gas in the well is static and the gas in the pipeline is moving.

Natural gas is bought, sold, and delivered on a continual basis, every day of the year, 24 hours a day. Careful, accurate measurement of the gas is very important. Properly calibrated meters ensure that these measurements are as accurate as possible. Meters are the cash registers of the pipeline industry. Pipeline personnel responsible for gas measurement use a variety of meters in their jobs, and they must keep these meters in good working order. Even small errors can cost companies many thousands of dollars.

Information about gas quality and measurement travels through type of pipeline—an administrative one. Pipeline personnel use computers to make accurate daily measurements and keep records. Because the functions of quality control and measurement overlap at times, clear communication also plays an important role.

2.0.0 ◆ SAMPLING AND TESTING

A pipeline company's ability to attract and keep customers depends on quality control. Natural gas may contain elements, such as carbon dioxide, hydrogen sulfide, nitrogen, and water vapor, that can affect quality. Quality control personnel take samples and run tests to check and maintain gas quality. Some of the elements for which natural gas is tested and the instruments used to do these tests are discussed in this section.

2.1.0 Dew Point

Dew point is the temperature at which condensation begins. Think of a glass filled with cold liquid. After a short time, the glass cools to almost the same temperature as the liquid. Once the glass becomes colder than the surrounding air, water vapor in the air cools and settles on the glass as small drops, or dew.

Moisture content in natural gas is measured using dew point and moisture meters, which are also called hygrometers. Hygrometers are used to measure the dew point in a range of temperatures. They are available in portable units for spot testing, in-line models, and models that allow continuous monitoring. There are both analog and digital versions. Some meters include a logging function that allows for storage of dew point measurements and other data.

2.2.0 Carbon Dioxide

Carbon dioxide (CO_2) is a heavy, colorless gas that occurs naturally in natural gas wells. CO_2 is removed from natural gas for several reasons:

- It reduces the heating value of natural gas.
- It is corrosive and can damage pipelines.
- It can freeze and form into clumps of dry ice, which may clog or damage pipeline equipment.

> **WARNING!**
> High CO_2 levels can cause paralysis in the lungs. Long periods of exposure to this gas may result in death. Personnel must wear the appropriate personal protective equipment when working in areas where CO_2 might displace oxygen.

2.3.0 Nitrogen

Nitrogen is colorless, odorless, tasteless, and nontoxic. However, its presence in natural gas lowers quality. The Gas Research Institute defines natural gas containing more than 4 percent nitrogen as subquality gas. For natural gas to be acceptable for shipment in most pipelines, the nitrogen must be removed. Pipeline personnel use nitrogen analyzers to detect this gas. Computer programs designed to work with the analyzer can produce a variety of reports that can be customized for each company.

2.4.0 Bacteria

Bacteria can form in pipes that carry natural gas or in any pipeline system where water can collect. The presence of bacteria can result in deposits in the pipeline that contain harmful acids. Bacteria in the pipeline can result in lower gas quality and stopped-up filters. Most pipeline companies use chemicals to clean out bacteria.

2.5.0 Hydrogen Sulfide

Hydrogen sulfide (H_2S) is a flammable, colorless compound of hydrogen and sulfur. It forms naturally when organic matter breaks down. It is found in crude oil, natural gas, sewers, and swamps. Although H_2S has a strong odor, it can deaden the sense of smell. Corrosive and poisonous, H_2S can paralyze your central nervous system. Inhalation can cause death.

> **WARNING!**
> Hydrogen sulfide is a very dangerous compound. It is corrosive and poisonous. Exposure can be deadly.

An H_2S analyzer is used to test for hydrogen sulfide. A pressure regulator in the unit controls the pressure and flow of the sample. A sensing tape reads the sample and records the concentration level of the gas. Readings are shown on a liquid crystal display. The layout of a hydrogen sulfide analyzer is shown in *Figure 1.*

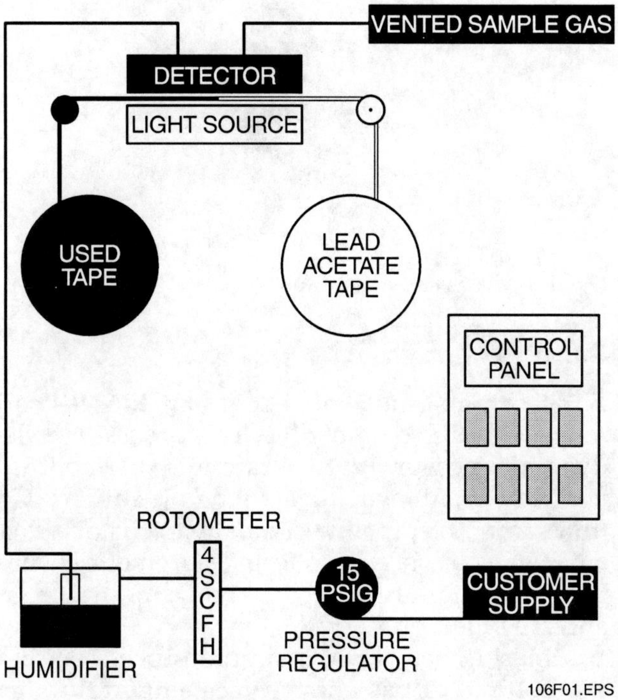

Figure 1 ◆ Hydrogen sulfide analyzer.

You must follow your company's safety policies when working near H_2S. You must also wear the appropriate personal protective equipment (PPE). Electronic sensors are available to detect toxic gases such as H_2S. Many of these sensors are easily maintained and calibrated in the field. Some models can be calibrated quickly by touching magnets to the correct areas on sensors. Available in both analog and digital models, many sensors also have built-in safety features that alert users if a sensor fails or is missing from the unit.

2.6.0 Moisture

It is important to keep excess water out of the gas stream. If water mixes with the CO_2 and H_2S found in natural gas, corrosive acids can form. These acids can damage the pipeline and its components. In addition, excess water condenses and forms ice crystals that can develop into plugs that block gas flow. Furthermore, too much moisture can reduce the amount of heat energy produced by natural gas. Pipeline personnel have several tools that help them detect moisture content. Two of those tools are moisture content analyzers and tube test kits.

2.6.1 Moisture Content Analyzer

Moisture content analyzers (see *Figure 2*) are computerized units that are available in both portable and stationary models. These units are contained in a housing that allows them to be used indoors or outdoors directly at the measurement point. A sensor in the unit absorbs moisture into a special film that covers two electrodes set in a glass tube. The moisture content is measured when an electric current passes through the electrodes. This current is proportional to the amount of absorbed moisture.

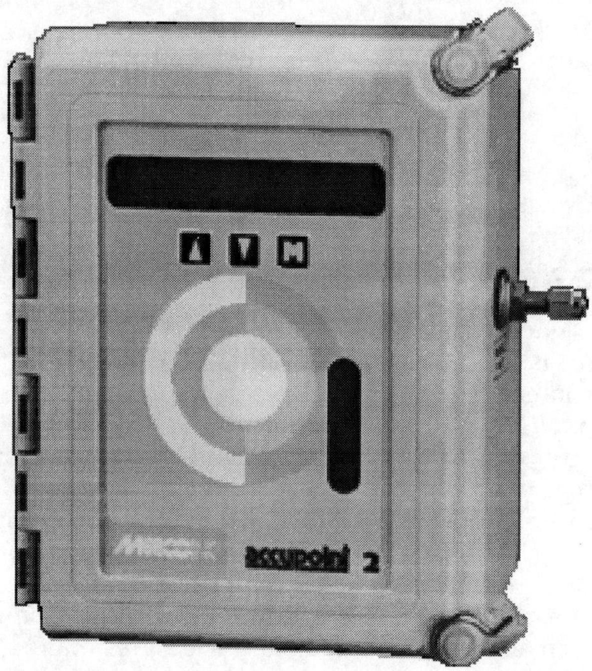

Figure 2 ◆ Moisture content analyzer.

2.6.2 Tube Test Kits

Pipeline personnel use tube test kits (see *Figure 3*) to test for moisture or hazardous gases such as H_2S. These tubes provide exact readings, can be operated with one hand, and are fairly easy to use. The operator simply inserts a tube into the unit. A small hand pump attached to the unit allows the operator to draw air into the tube. Calibrated marks on the tube show the exact level of the gas or moisture.

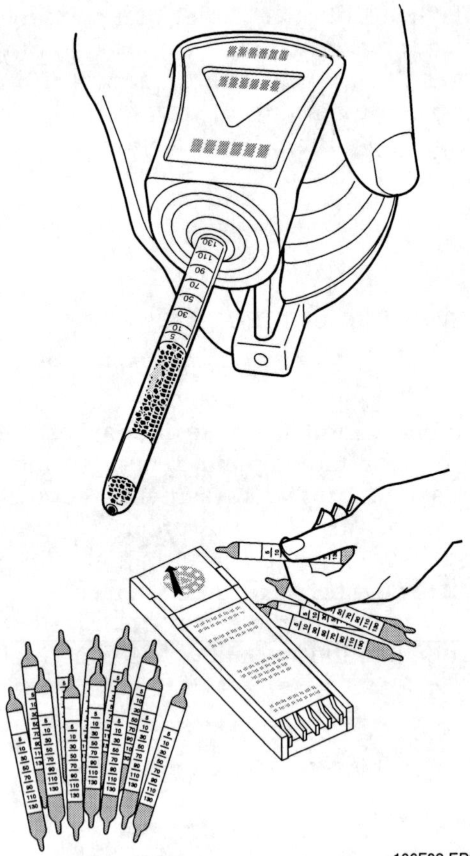

106F03.EPS

Figure 3 ◆ Typical tube test unit with a selection of tubes.

2.7.0 Gas Chromatograph

Gas chromatographs (see *Figure 4*) are used to measure heating values and liquid or gas vapor content. They also can detect the percentages of oxygen, nitrogen, carbon dioxide, and carbon monoxide that may be in the sample. Gas chromatographs may be used at the pipeline facility or samples may be sent to a laboratory for analysis.

Gas chromatographs come in two basic types: British thermal unit (Btu) and process. The Btu chromatograph is used in natural gas transmission and distribution lines. It measures many gas quality values, including Btu content, relative density, and compressibility. The process chromatograph is used in natural gas pipelines and processing operations. This unit can analyze liquid or gas vapor for traces of sulfur and hydrocarbons. Control systems that are used to manage these instruments are also available.

The gas chromatograph works as follows. A natural gas sample is placed into a stream of another gas, usually helium or argon. This gas is called the carrier. The carrier takes the sample through a column packed with a solid material that has been finely divided. These divisions allow the carrier and sample to pass through the material. Elements in the sample move through the column at different speeds depending on how they react to the material in the column. The slower-moving elements separate from the faster-moving elements. Each element then passes through a detector, where each is analyzed. Test results are recorded on a graph.

106F04.EPS

Figure 4 ◆ Gas chromatograph.

3.0.0 ◆ METERS

Pipeline personnel must measure the gas as it moves through the pipeline. This process is called dynamic measuring. Accurate measurement is a very important part of pipeline operations. Gas flows from the pipeline company to customers on a constant basis. Accurate measurement ensures that customers receive the right amount of gas and are billed properly.

Metering gas is not a simple process because the volume of gas changes as temperature and pressure change. Therefore, carefully calibrated meters are important tools in the measurement of gas. Three basic types of meters are used: turbine, orifice, and ultrasonic flow. Meters must be calibrated before use and checked on a regular basis to ensure accurate measurement.

3.1.0 Turbine Meters

In the gas turbine meter (see *Figure 5*), the fan-like turbine rotates at a speed that corresponds to the speed of the passing element (fluid or gas). In fact, it is the speed of the passing element, not its volume, that makes the turbine meter work. Turbine meters must be carefully installed to ensure a straight flow. To properly install the meter, workers use a straight pipe the same size as the meter. To ensure the meter's accuracy, manufacturers recommend that this pipe reach at least 10 pipe diameters upstream and at least 5 pipe diameters downstream (see *Figure 6*).

106F05.EPS

Figure 5 ◆ Gas turbine meter.

Turbine meters measure speed and convert it to volume. Rotors in the meter are fitted with a signaling device. The meter makes one complete turn and sends a signal to an electronic counter. The total of the signals is divided by the frequency per unit of volume.

3.2.0 Orifice Meters

The plate (see *Figure 7*) is the most important part of the orifice meter. The plate restricts flow and produces a difference in pressure. This pressure difference is measured to calculate actual flow. Expert installation and maintenance ensure the meter's accuracy.

Orifice plates may be fastened between the halves or flanges of machined carriers. They may also be placed in the body of a meter housing, which allows workers to easily remove the plate to check it. On some meters, the plate may be removed without stopping the flow. A typical orifice meter includes a carrier housing or flanges, a

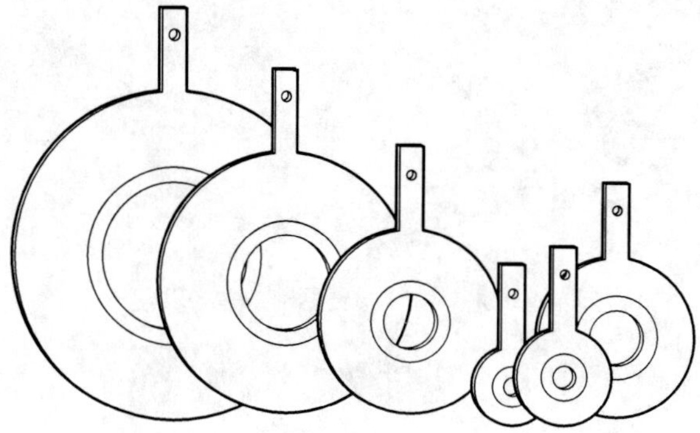

106F07.EPS

Figure 7 ◆ Sharp-edged, concentric orifice plates.

plate, gaskets and O-rings, removal devices, and pressure fittings. Some common orifice plate holders are shown in *Figure 8*. To prevent leaks, workers must do the following tasks:

- Check the housing unit and the housing joints for corrosion or cracks
- Check the pressure fittings and the sensor connections for leaks or damage
- Ensure that the plate is the right thickness, polished, smooth, and straight

NOTE

Orifice plates must be of the correct thickness. Workers use a micrometer, which is designed for very small measurements, to accurately check plate thickness.

- Check all the bolts and fittings for cracks
- Replace worn or damaged O-rings and gaskets
- Clean off any dirt
- Ensure that the orifice plate is the right size for the job
- Check all measurements
- Document all measurements

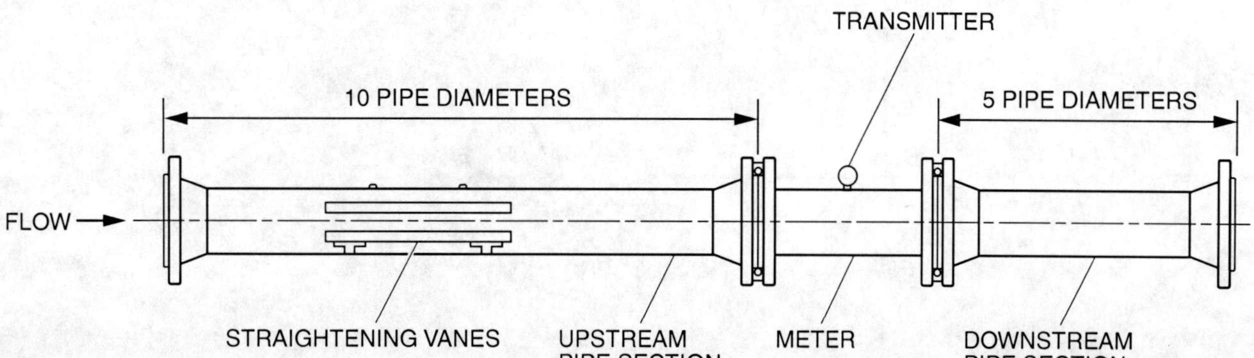

106F06.EPS

Figure 6 ◆ Typical meter run layout.

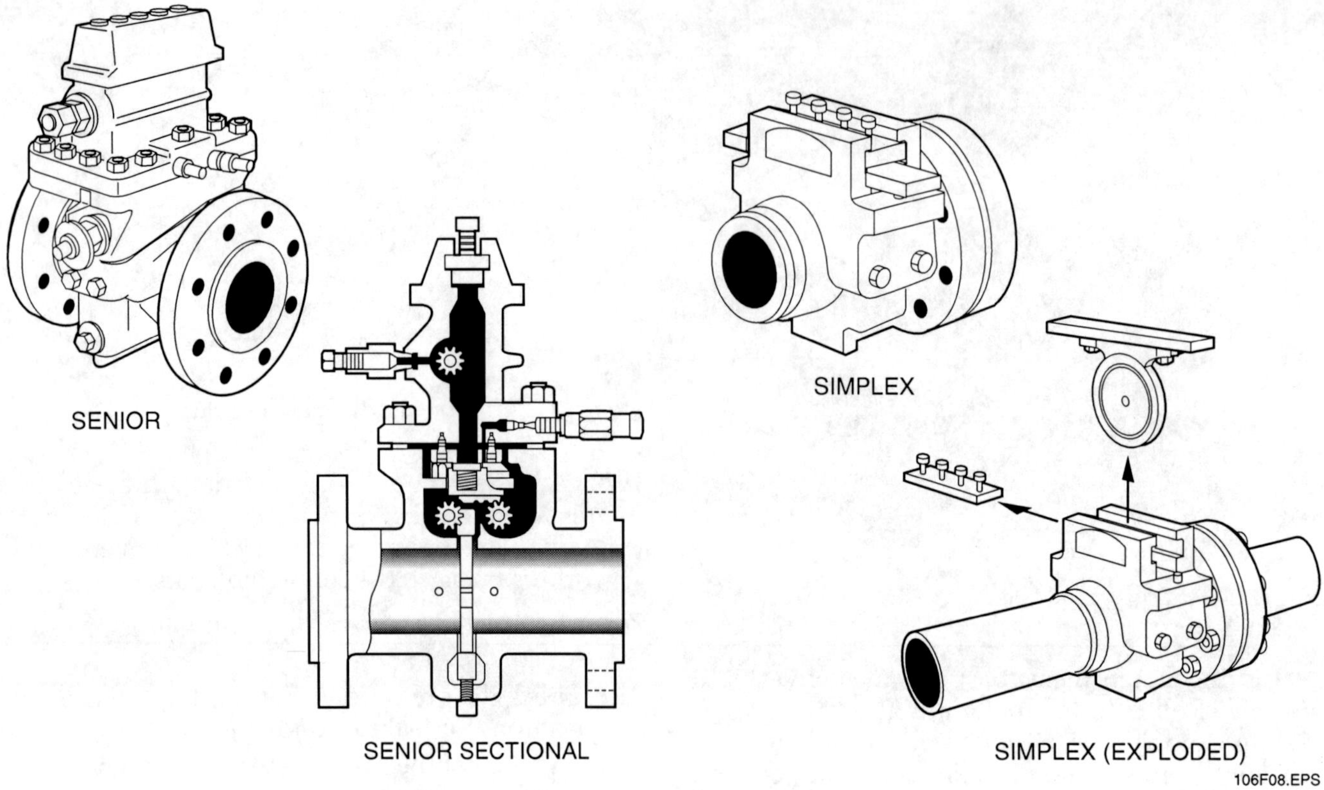

Figure 8 ◆ Common orifice plate holders.

3.3.0 Ultrasonic Flow Meters

An ultrasonic flow meter (see *Figure 9*) is also known as a transit-time meter because it measures the amount of time it takes for an ultrasonic sound wave to transit, or pass, from one sensor to another. This sound wave is sent and received by sensors called **transducers.** The transit time is measured in both the upstream and downstream directions. Because the sound wave is carried with the gas flow, the downstream transit time is shorter than the upstream transit time. Ultrasonic flow meters come in two basic types: single-path and multipath.

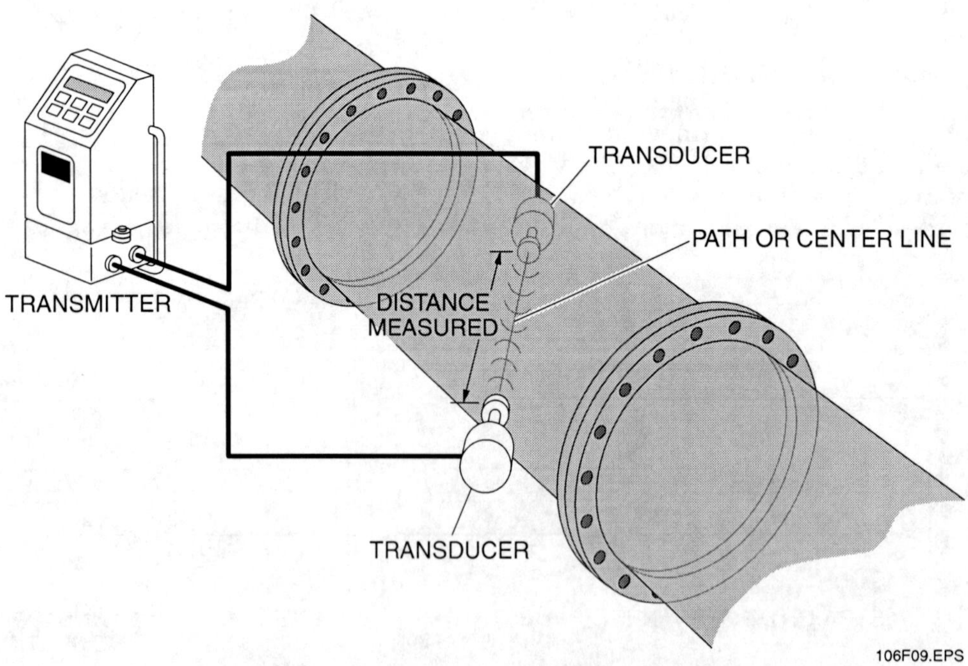

Figure 9 ◆ Ultrasonic flow meter.

In a single-path meter, the transit time is measured with a single pair of transducers. The transit time (stated in feet per second) is calculated by dividing distance by time. To compute flow rate, the meter compares the difference in travel rate with the speed of sound for known gas properties. The sound waves' rate of speed is affected by temperature and pressure.

A multipath ultrasonic meter works exactly like a single-path meter. However, the multipath meter uses several pairs of transducers to send and receive signals. The meter records the transit time for each path and calculates average flow speed. Because more information is collected, multipath readings tend to be more accurate than single-path readings.

3.4.0 Inspecting Meters and Calibrating Transmitters

Meters and the transmitters that send information from them to the control center must be inspected on a regular basis to ensure accuracy. Meters and transmitters are calibrated at the factory where they are made. Over time, however, readings can start to vary. Therefore, meters must be inspected and, if necessary, their transmitters must be recalibrated. Calibration is the process of setting a meter to a known standard. Recalibration is the process of making adjustments to meters that vary from that standard.

Meter inspection and transmitter calibration are so important in the pipeline industry that some customers ask to **witness** the process. Many companies notify their customers when this process is scheduled and invite them to watch. During this process, pipeline personnel test the meters under conditions in which flow rates, temperature, and pressure are as close as possible to actual operating conditions.

3.5.0 Strainers

Meter efficiency is affected by contaminants in the gas stream. To ensure a clean gas stream, strainers are used. Strainers come in several types in order to handle different applications. For example, some strainers are designed to handle small amounts of particles in the gas stream. In-line strainers have finer mesh screens and can handle larger dirt loads. Process strainers have the highest dirt-loading capacity, which means that they are able to trap the highest amount of contaminants. Strainers are used for several purposes:

- To clean the gas stream and increase gas quality
- To protect gas pipeline equipment from contaminants
- To reduce maintenance on downstream equipment
- To extend equipment life

Pipeline personnel must inspect strainers regularly to clean or replace the mesh elements that trap particles. It is best to follow the manufacturers' instructions for the installation, use, and maintenance of strainers.

3.6.0 Remote Terminal Units

Remote terminal units (RTUs) act like message centers for pipeline operations. They process information collected from equipment such as meters and send and receive signals on a continual basis. RTUs are part of the pipeline's Supervisory Control and Data Acquisition (SCADA) computer system. Transmitters in the SCADA system send signals to the RTU. Following are examples of some of the types of information sent by RTUs:

- Readings of measurable variables recorded by meters
- Readings of logical operations in the computer system
- A frequency signal that can be compared with a measurable variable

The RTU, in turn, sends signals to actuators along the line. Three main types of signals are sent by an RTU:

- Signals sent to actuators that control pressure
- Signals that trigger mechanical actions, such as opening or closing a valve
- Signals sent to other RTUS or controls along the pipeline

Pipeline personnel responsible for working with these units must monitor their accuracy and effectiveness on a regular basis. Follow the manufacturer's guidelines and company policy when working with these units.

4.0.0 ◆ DAILY MEASUREMENT STATEMENTS

Because natural gas is delivered around the clock, daily measurements are essential. Accurate statements are key to all of the administrative work on the pipeline, such as orders, billings, receipts, and payments. Following are some of the tasks that measurement personnel perform to ensure the accuracy of daily statements:

- Ensure that all custody transfer equipment is checked for proper calibration and operation
- Accurately measure the gas before it is sent

- Monitor progress of the gas through the pipeline and at custody transfer stations
- Ensure that the correct amount of gas has been sent
- Prepare accurate, itemized daily measurement statements

During transmission of large volumes of gas, a little more or a little less gas may be delivered. Pipeline personnel use the SCADA system to monitor these overages and shortages to ensure that they remain within accepted industry standards and to adjust billing if necessary.

5.0.0 ◆ MONITORING ODORANT LEVEL

During the early years of its use, natural gas was not odorized. Many thought that the sulfur naturally found in the gas could provide sufficient warning of leaks. However, the sulfur levels were often too low to be easily detected. Although research on odorants was being done, finding a safe, reliable odorant was not a simple process. In 1937, natural gas leaked into a school in New London, Texas. Because the gas was odorless, it accumulated undetected. Then a spark from an electric tool in a shop class ignited an explosion that flattened the school. Many teachers and students were killed, and the tragedy led to the first laws to regulate gas odorization.

5.1.0 Odorization Requirements

Today, federal law requires that odorant be added to natural gas so that a person with a normal sense of smell can detect a leak. However, not all gas pipelines require odorization. Distribution lines must be odorized. Generally, it is the responsibility of the local gas distributor to inject and monitor odorant on these lines. Transmission lines, offshore lines, and lines that run through thinly populated areas are generally not required to contain odorant.

The Code of Federal Regulations (*49 CFR 192.625*) states when and where odorant must be added to natural gas. Pipeline companies must follow this law when sending combustible gas through certain populated areas located near the pipeline. The law divides areas located near pipelines into **class location units.** A class location unit is an onshore area that extends 220 yards (200 meters) on either side of the centerline of any continuous 1-mile (1.6-kilometer) length of pipeline. There are four class location units (see *Table 1*). The odorization requirement applies mainly to Classes 3 and 4.

Table 1 Class Location Units

Class Number	Definition
Class 1	An offshore area or an area that has 10 or fewer buildings in which people live.
Class 2	An area that has more than 10 but fewer than 46 buildings in which people live.
Class 3	An area where there are 46 or more buildings in which people live. An area where the pipeline lies within 100 yards (91 meters) of a building. An area where the pipeline lies within 100 yards (91 meters) of a well-defined outdoor space that is occupied by 20 or more people at least 5 days a week for 10 weeks in any 12-month period. (The days and weeks do not have to be consecutive.) Examples of such spaces include playgrounds, recreation areas, and outdoor theaters.
Class 4	An area containing several buildings that are four or more stories aboveground.

Source: Code of Federal Regulations *49 CFR 192.5, Class*

The regulation allows for some exceptions. Natural gas does not have to be odorized in a distribution line in a Class 3 or a Class 4 location under certain conditions. For a full discussion of those conditions, it is best to consult *49 CFR 192.625.*

The right type and amount of odorant must be added to combustible gases. Odorant is injected into the system using a computer program that monitors the levels and sounds an alarm if proper levels are exceeded. Odorant must meet the following safety conditions:

- The odorant must not be harmful to people, property, or the environment.
- Combustion products from the odorant must not be toxic when breathed.
- Combustion products from the odorant must not corrode or harm materials they touch.
- The odorant must not dissolve in water to an amount greater than 2.5 parts to 100 parts by weight.

Odorant must be added to combustible gases in the right concentration, or amount. It is not acceptable to have wide variations in the level of odorant. Companies are required by *CFR 192.625* to conduct periodic sampling to ensure the proper concentration of odorant. Note that operators of master meter systems may meet this requirement by getting a notice in writing from their gas source that the gas has the right amount of odorant and by performing regular sniff tests to ensure that the gas contains odorant.

5.2.0 Safety Considerations

Most odorants contain mercaptan, a compound that contains sulfur. Mercaptan produces a rotten-egg smell that is easily recognized. When injecting odorants, workers must read the manufacturer's material safety data sheet (MSDS) and follow all of the manufacturer's safety recommendations. In general, the MSDS provides the following information:

- Hazards to people coming into contact with the product
- Hazards to the environment
- Chemical composition of the product
- Fire or explosion hazards
- Potential long-term health effects
- First aid and emergency instructions
- Recommendations for the appropriate PPE

6.0.0 ◆ LEAKAGE SURVEYS

Leakage surveys on gas transmission pipelines must be performed on an ongoing basis according to *49 CFR 192.706.* A leakage survey must be performed at regular intervals not exceeding 15 months, but at least once each calendar year. The intervals depend on the type of location:

- In Class 3 locations, at intervals not exceeding 7½ months, but at least twice each calendar year
- In Class 4 locations, at intervals not exceeding 4½ months, but at least four times each calendar year

An unexpected release of gas is defined as an abnormal operating condition. All responsible workers must be prepared to deal with an unexpected gas release. In order to do this, they must be familiar with the following:

- Company policies and safety procedures
- Government regulations
- Company procedures for responding to emergencies

The proper procedure for dealing with a gas leak depends on your company. All companies have their own detailed plans for emergencies. These plans include all the steps employees must follow and the order in which they must follow them. A company's plan for handling gas leaks most likely includes instructions for the following procedures:

- Evaluating the problem
- Finding and safely isolating the source of the leak
- Informing the appropriate company personnel
- Notifying the local emergency response agencies if appropriate
- Helping to coordinate a general evacuation plan if necessary

> **NOTE**
> In the event of a major emergency, an evacuation of the general public may be necessary. Pipeline companies do not have the authority to order such an evacuation. Only the local emergency response authorities may do so. Pipeline companies work with the appropriate agencies and give them the information they need to decide whether a general evacuation is required.

The federal regulations that govern leakage surveys of gas distribution lines are different from those governing transmission lines. For details consult *49 CFR 192.723.*

6.1.0 Line Inspection

Pipeline workers may use a hydrogen flame ionization unit to detect gas leakage along gas transmission pipelines (see *Figure 10*). This unit is based on a simple scientific principle. A hydrogen flame does not conduct electricity, but the hydrocarbons in natural gas do. When gas is present, an electric current passes between two electrodes in the unit. The unit measures the current to give a reading of gas levels in the air. This type of unit works best in detecting low concentrations of gas (below 100 parts per million).

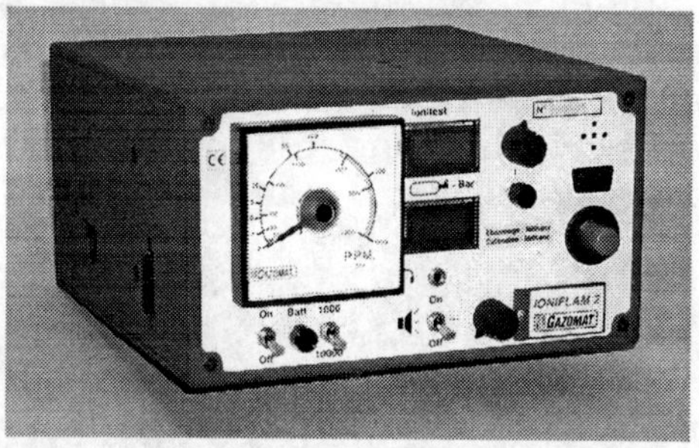

106F10.EPS

Figure 10 ◆ Hydrogen flame ionization unit.

Pipeline operators must have a patrol program in place to check for leaks, construction activity, or any other conditions affecting the safety and operation of the pipeline. Patrol frequency depends on several factors:

- The location of the pipeline
- The size of the pipeline
- The operating pressures
- The weather

6.2.0 Patrol Intervals

Maximum regulated intervals between patrols for gas pipelines are shown in *Table 2*. Your company's policies state the actual patrol schedule for a given pipeline. Personnel may patrol the pipeline by walking, driving, or flying.

Table 2 Maximum Interval between Patrols

Class Location of Line	Maximum Interval between Patrols	
	At Highway and Railroad Crossings	At All Other Places
1 and 2	7½ months, but at least twice each calendar year	15 months, but at least once each calendar year
3	4½ months, but at least four times each calendar year	7½ months, but at least twice each calendar year
4	4½ months, but at least four times each calendar year	4½ months, but at least four times each calendar year

Summary

Pipeline companies could not operate without quality control and accurate measurement. Customers want natural gas that is clean and free of harmful elements. The government sets safety standards that must be met. Therefore, quality control personnel must test samples to ensure that the gas meets all standards for cleanliness and safety. Measurement personnel help to ensure that there is enough gas to meet demand. Their work is also important in customer service. Customers do not want to pay for products they have not received, so accurate measurement is key to customer satisfaction. Finally, because even small errors can cost many thousands of dollars, measurement plays an important role in a pipeline company's financial health.

Review Questions

1. H_2S is a flammable, colorless compound of _____.
   a. hydrogen and selenium
   b. hydrogen and sulfur
   c. helium and strontium
   d. helium and sulfur

2. Pipeline personnel use tube test kits to test for _____.
   a. pressure differences in the line
   b. sediment and water that might be present in the gas
   c. moisture and hazardous gases
   d. differences in flow rate and temperature

3. Turbine meters measure speed and convert it to _____.
   a. temperature
   b. pressure
   c. volume
   d. flow

4. The _____ is the most important part of the orifice meter.
   a. micrometer
   b. flange
   c. housing
   d. plate

5. In an ultrasonic flow meter, transducers _____.
   a. send and receive sound waves
   b. send and receive electromagnetic impulses
   c. send but do not receive sound waves
   d. send but do not receive electromagnetic impulses

6. The Code of Federal Regulations (*49 CFR 192.65*) defines areas located near pipelines by dividing them into _____ units.
   a. right-of-way
   b. population count
   c. pipeline grid
   d. class location

7. Most odorants are based on mercaptan, a compound that contains _____.
   a. hydrogen sulfide
   b. sulfur
   c. mercury and capsicum
   d. phosphorous

8. *49 CFR 192.706* requires that leakage surveys be performed on gas transmission pipelines at regular intervals _____.
   a. not exceeding 15 months, but at least once each calendar year
   b. not exceeding 18 months, but at least once each calendar year
   c. according to company policy, but at least once each calendar year
   d. according to company policy, but at least once each 18-month period

9. An unexpected release of gas is defined as a(n) _____.
   a. abnormal emergency event
   b. abnormal operating condition
   c. condition requiring U.S. Coast Guard intervention
   d. condition requiring immediate shutdown of all systems

10. In the event of a major emergency, only the _____ is (are) authorized to order an evacuation of the general public.
    a. company or companies that own the pipeline
    b. owner of the land where the pipeline is located
    c. local emergency response authorities
    d. mayor and town council

GLOSSARY

Trade Terms Introduced in This Module

Class location unit: An area defined by the Department of Transportation in the Code of Federal Regulations as an onshore area that extends 220 yards (200 meters) on either side of the center line of any continuous 1-mile (1.6-kilometer) length of pipeline.

Transducer: In an ultrasonic flow meter, any of a pair of units that send and receive ultrasonic sound waves.

Witness: In the pipeline industry, to observe the process that ensures that equipment is calibrated properly and making accurate measurements.

Figure Credits

Meeco, Inc.	106F02
Emerson Process Management, Daniel Division	106F04, 106F05
TD Williamson, Inc.	106F10

CONTREN™ LEARNING SERIES — USER UPDATES

The NCCER makes every effort to keep these textbooks up-to-date and free of technical errors. We appreciate your help in this process. If you have an idea for improving this textbook, or if you find an error, a typographical mistake, or an inaccuracy in NCCER's Contren™ textbooks, please write us, using this form or a photocopy. Be sure to include the exact module number, page number, a detailed description, and the correction, if applicable. Your input will be brought to the attention of the Technical Review Committee. Thank you for your assistance.

Instructors – If you found that additional materials were necessary in order to teach this module effectively, please let us know so that we may include them in the Equipment/Materials list in the Instructor's Guide.

Write: Curriculum Revision and Development Department
National Center for Construction Education and Research
P.O. Box 141104, Gainesville, FL 32614-1104

Fax: 352-334-0932

E-mail: curriculum@nccer.org

Craft ______________________ Module Name ______________________

Copyright Date __________ Module Number __________ Page Number(s) __________

Description

____

____

____

____

(Optional) Correction

____

____

____

(Optional) Your Name and Address

____

____

____

Module 67107-02

Abnormal Operating Conditions

Introduction to the Pipeline Operator Qualification Rule, API, and NCCER

It could be said that petroleum products keep America running. Petroleum products like oil and natural gas are the principal sources of energy in the United States and are used in many manufacturing and chemical processes. Petroleum products are primarily transported by pipeline. There are more than a million miles of liquid and natural gas pipeline in this country, ranging in diameter from two inches to four feet. These pipelines bring crude oil and natural gas from production fields to processing plants and refineries—and ultimately to the consumer.

Pipeline Operator Qualification

Petroleum products are considered hazardous, so pipeline personnel must take extra precautions and follow specific procedures when working with them. To help ensure the safe handling and transportation of petroleum products, the United States Department of Transportation (DOT) mandates safe practices as federal regulations.

One such regulation, the Operator Qualification (OQ) rule, requires pipeline operators to develop a written qualification plan to evaluate their personnel and their contract personnel (see *49 CFR Part 192 Subpart N—Gas,* and *49 CFR Part 195 Subpart G—Liquid*). The evaluation hinges on the operator's ability to perform certain covered tasks and to be able to recognize and respond to abnormal operating conditions that may be encountered while performing these covered tasks.

Covered Tasks

A pipeline performance task is deemed *covered* if the activity passes the following four-part test:

1. Is it performed on a pipeline facility?
2. Is it an operations or maintenance task?
3. Is it performed as a requirement of *49 CFR Part 195 (Liquid)* or *49 CFR Part 192 (Gas)*?
4. Does it affect the operation or integrity of the pipeline facility?

If the answer to all four questions is *yes*, then the activity is considered *covered* and must figure into the evaluation process of pipeline personnel.

American Petroleum Institute

The OQ rule and its related covered tasks provided the catalyst to develop a unified pipeline training and assessment program—so that pipeline operators and owners have ready access to the assessment and training tools they need. This effort was the direct result of the vision and leadership provided by The American Petroleum Institute (API) and API members.

Efforts to unify the industry are nothing new to API. Indeed, the impetus for forming the Institute in 1919 was the need to standardize engineering specifications for drilling and production equipment. Today, the most pressing issues revolve around public perceptions and government policies toward the industry. Speaking with one voice on these issues has become as essential as having interchangeable parts in the field. API's members determine the petroleum industry's positions on public policy as well as the standards that govern its day-to-day operations worldwide.

Consortium on Operator Qualification

API assembled a focus group whose primary purpose was to identify covered tasks and provide compliance recommendations to pipeline operators.

The Consortium on Operator Qualification (COOQ) became the guiding force for the identification and development of covered tasks for the liquid pipeline industry. Their work product, API publication 1161, *Guidance Document for the Qualification of Liquid Pipeline Personnel*, formed the basis for the resulting Pipeliner Training and Assessment Program.

The covered task list created in API publication 1161 has now evolved to include gas covered tasks and is maintained by NCCER. The most current covered task list can be downloaded at www.nccer.org.

National Center for Construction Education and Research

With the guidance and support of API, its members, and the Consortium on Operator Qualification, the National Center for Construction Education and Research (NCCER) was able to design and implement the Pipeliner Training and Assessment Program.

NCCER is a not-for-profit 501(c)(3) education foundation established in 1995 by the world's largest and most progressive construction and maintenance contractors and national contractor associations. NCCER was founded to address the

severe workforce shortage facing the construction and maintenance industries and to develop a standardized training process and curricula. Today, NCCER is supported by hundreds of leading construction and maintenance companies, manufacturers, and national associations and developed more than 35 construction and maintenance titles.

NCCER also maintains a National Registry, which contains the records of trainees and instructors who have received NCCER training. All completion information for every trainee is sent to the NCCER and filed within the Registry. The Registry can then confirm training and skills for workers as they move from company to company, state to state, or even within their own company.

Standardized Curriculum (NCCER's Contren™ Learning Series)

NCCER curricula are developed using the expertise of industry subject matter experts (SMEs). The use of SMEs as an integral part of every new development or revision ensures that the curriculum is current, industry-driven, and indicative of industry practices. The development and/or revision process is facilitated from start to finish by NCCER project managers, who work closely with the SMEs and a select team of technical writer(s) on the development and presentation of the curriculum.

Pipeline: Five Training Paths

The pipeline curriculum focuses on essential skills and competencies within five skill areas:

- Corrosion Control
- Electrical and Instrumentation
- Maintenance
- Mechanical
- Operations (Field and Control Center)

SMEs from the pipeline industry have determined the skills and competencies needed based on API's covered tasks and the consensus of the group/industry. Specific company information is generally not incorporated into the curriculum unless the SMEs determine that the information is the industry standard.

Training Modules and Covered Tasks

The smallest unit of instruction in the curriculum is a *module*. Each curriculum is composed of modules (the equivalent of stand-alone chapters) that are logically sequenced into complete *levels* of instruction. Every module is task-oriented and requires the successful completion of a written (knowledge-based) examination and a performance (hands-on) test. Modules may be individually selected and/or combined for efficient task training. Modules in the pipeline curricula, to the extent possible, will align with API-developed covered tasks. While not all modules will relate to covered tasks, all covered tasks will relate to at least one module.

Benefits of Training to the Pipeline Marketplace

Applying NCCER's standardized training process allows pipeline owners and contractors in the United States to establish the benchmark for acceptable skill and competency levels within the pipeline industry. This process will assist training organizations in developing and/or improving standardized pipeline training programs that effectively match their resources and operating conditions while responding to regulatory demands and to the needs of the industry. The process will also recognize training organizations that meet the minimum standards of competence, warranting regulatory, public, and professional confidence. This in turn will foster national unity within the pipeline industry for the mutual benefit of individual technicians, owners, contractors, and ultimately pipeline-users and society.

Additional Resources

To review the Pipeliner Training and Assessment Program Implementation and Resource Guide and the latest version of pipeline covered tasks, visit NCCER's Web site at www.nccer.org, or contact:

National Center for Construction Education and Research
PO Box 141104
Gainesville, FL 32614-1104
(888) 622-3720
Web site: www.nccer.org

For more information on API, please contact:

American Petroleum Institute
1220 L Street, Northwest
Washington, D.C. 20005
(202) 682-8000
Web site: www.api.org

To order *API Pub 1161*, contact Global Engineering Documents by calling 1-800-854-7149 or visiting their Web site at www.global.his.com.

Additional sources of information on the pipeline operator qualification regulation include:

DOT – Office of Pipeline Safety
www.ops.dot.gov

Transportation Safety Institute
www.tsi.dot.gov

COURSE MAP

This course map shows all of the modules in the Gas Pipeline Operations curriculum. The suggested training order begins at the bottom and proceeds up. Skill levels increase as you advance on the course map. The local Training Program Sponsor may adjust the training order.

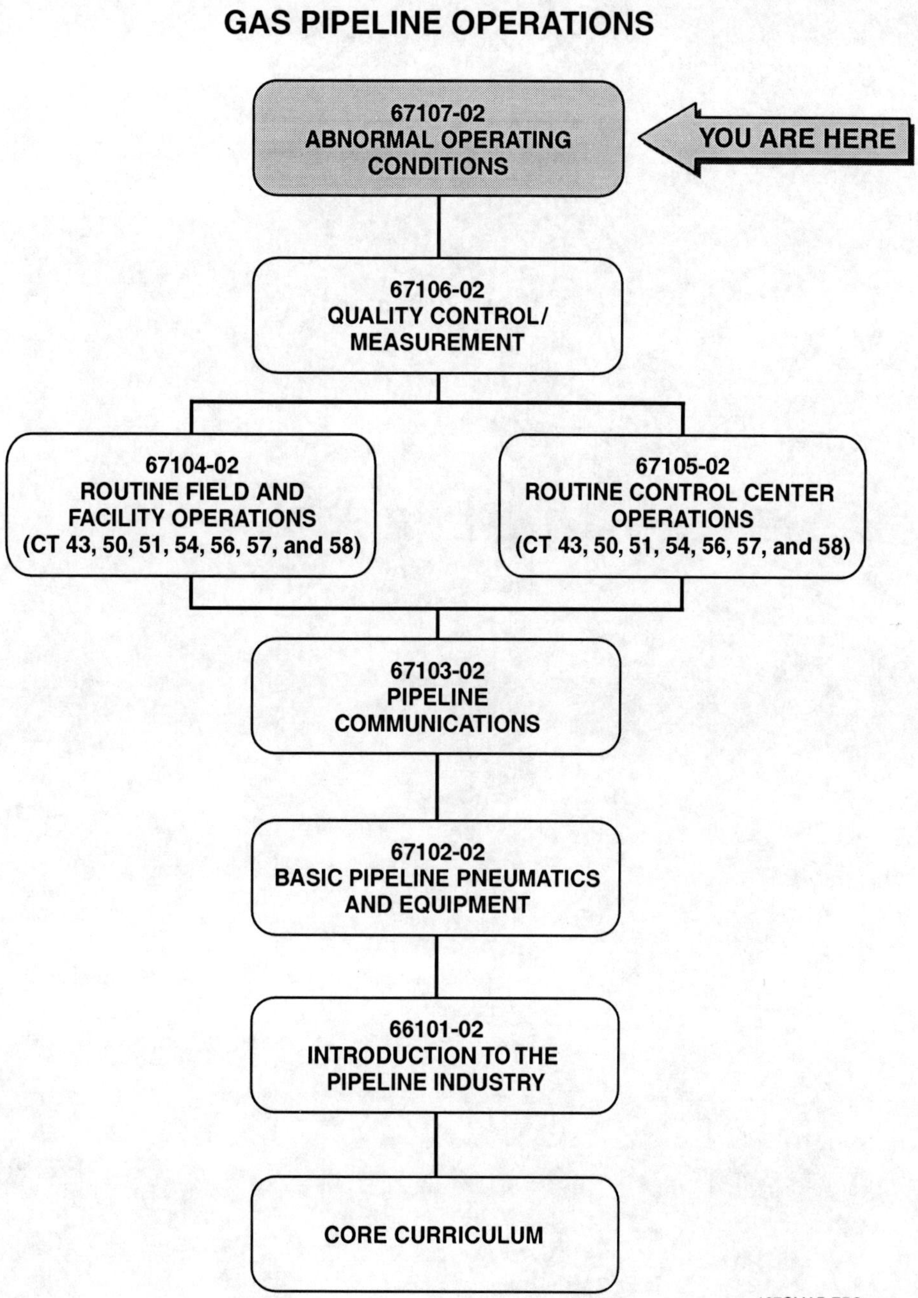

107CMAP.EPS

MODULE 67107-02 CONTENTS

Figures

Tables

MODULE 67107-02

Abnormal Operating Conditions

Objectives

When you have completed this module, you will be able to do the following:

1. Recognize and react to abnormal facility conditions.
2. Recognize and react to the activation of a safety device.
3. Recognize and react to communications failures.
4. Recognize and react to control system failures.
5. Recognize and react to power interruptions.
6. Recognize and react to fire, explosions, or natural disaster.
7. Recognize and react to pipeline system damage.
8. Recognize and react to unexpected release of a hazardous gas.
9. Recognize and react to unexplained pressure changes.

Prerequisites

Before you begin this module, it is recommended that you successfully complete the following modules: Core Curriculum; Pipeline Core, Module 66101-02; Gas Pipeline Operations, Modules 67102-02 and 67103-02, 67104-02 or 67105-02, and 67106-02.

1.0.0 ◆ INTRODUCTION

To ensure safety, federal laws require operators of transmission pipelines to have a manual for operations, maintenance, and emergencies (*49 CFR 192.605*). The manual must include information to help workers identify and respond to abnormal operating conditions (AOCs). An AOC is defined as any of the following:

- Any condition that may point to the failure of a system component
- Any condition that is a change from normal operations
- Any condition that may go beyond design limits
- Any condition that may harm people, damage property, or harm the environment

Types of AOCs, along with examples and responses, are listed in *Table 1*. Note that the responses are general. Each pipeline company has specific procedures to follow and specific reports to complete regarding AOCs. Therefore, you must know and follow your company's procedures. Those procedures are especially important when it comes to deciding whether to shut in a segment of the pipeline or the facility.

AOCs are a common occurrence, and pipeline companies have developed effective procedures and policies to enable employees to respond quickly, safely, and effectively. Most pipeline equipment is fitted with safety devices to alert personnel to problems. In addition, computer monitors in the control center help keep small problems from turning into big ones. In some cases, an AOC is a routine matter that can be quickly and safely addressed by personnel. In other cases, an AOC can result in an emergency situation. In all cases, pipeline personnel must follow established company procedures when responding to an AOC.

Table 1 Abnormal Operating Conditions: Examples and General Responses

Abnormal Operating Condition	Examples	General Responses
Abnormal facility condition	• Uncovered pipeline that should be covered or damaged pipeline • Damaged or missing pipeline supports • Corrosion • Coatings separating from the pipeline (disbondment) • Missing signs or markers	Make notifications per company policy. Complete repairs as necessary. Complete the required reports.
Control system failure	• Control system component failure • Pneumatic system failure	Follow company troubleshooting procedures. Switch to an alternate control system or power source if necessary. Make notifications per company policy. Complete the required reports.
Power interruption	• Transformer failure • Downed power lines	Follow company troubleshooting procedures. Switch to an alternate control system or power source if necessary. Make notifications per company policy. Complete the required reports.
Fire, explosion, natural disaster	• Electrical fire • Ignition of a hazardous vapor • Unintentional detonation • Lightning strike • Activation of a smoke or fire detector • Extreme weather (tornado, earthquake, hurricane) • Pipeline movement caused by an earthquake, landslide, or flood	Shut in the pipeline segment or facility if necessary. Evacuate facility if necessary. Make notifications per company policy. Complete the required reports.
Activation of a safety device	• Pressure relief valve activates • Emergency or abnormal shutdown of a device • High-pressure shutdown device • Hazardous air alarm	Shut in the pipeline segment or facility if necessary. Make notifications per company policy. Complete the required reports.
Pipeline system damage	• Damaged or inoperable components (valves, compressors, meters) • Accidental or intentional damage to the pipeline	Make notifications per company policy. Shut in the pipeline segment or facility if necessary. Complete the required reports.
Unexpected hazardous gas	• Hydrogen sulfide • Carbon dioxide	Follow troubleshooting procedures. Shut in the pipeline segment or facility if necessary. Make notifications per company policy. Complete the required reports.
Unexplained flow rate change	• High flow • Low flow • No flow	Follow troubleshooting procedures. Shut in the pipeline segment or facility if necessary. Make notifications per company policy. Complete the required reports.
Unexplained pressure change	• Increased pressure • Decreased pressure • No pressure	Follow troubleshooting procedures. Shut in the pipeline segment or facility if necessary. Make notifications per company policy. Complete the required reports.

In general, employee response can be divided into five steps:

- Recognizing or identifying the AOC
- Carrying out company safety procedures to protect people, property, and the environment
- Investigating the cause or causes of the AOC
- Making repairs if appropriate
- Completing the required reports

Reports are required by your company, and some reports are required by state and federal regulatory agencies. The number of laws and company rules can seem overwhelming at times. However, it is important to remember that these laws and rules protect the safety of employees, the public, and the environment. They are also designed to limit or prevent damage to pipeline equipment. Many company rules are based on the laws that govern pipeline operations, so it is important to always follow the rules and get the proper training. When an AOC occurs, you must know what to do. If you learn about AOCs and plan the correct response before a problem occurs, you help keep pipeline operations safe for everyone.

Completing the required reports is an important part of responding to AOCs. Your company must keep a record of what happened. You must document the day, time, and all actions you took during an AOC. Government agencies also need information about AOCs. They use this information to keep track of pipeline safety issues. A number of organizations, both private and government (local, state, and federal), track information about the pipeline industry. A list of some of these organizations is in *Appendix A*.

2.0.0 ◆ REQUIRED DOCUMENTS AND REPORTS

Federal law (*49 CFR 191.23*) requires pipeline companies to send written reports to the Office of Pipeline Safety in Washington, D.C. Companies submit information about an AOC on a safety-related condition report (see *Figure 1*). An incident report is used for company files (see *Figure 2*).

Companies must keep all documents related to AOCs on file. A complete file includes work orders, permits, and any lockout tags (see *Figure 3*) placed on equipment. If the company conducts an investigation, the file must also include the investigator's notes and photographs. Authorized personnel must sign and initial all documents.

2.1.0 Post-AOC Investigation

Following some types of AOCs, the company may be required by law to conduct an investigation. Investigations also serve a practical purpose. They help management and employees learn why the problem happened and find ways to keep it from happening again if possible. For example, if the cause is human error, the investigator may recommend additional training. The investigator should put together a file that may include some or all of the following items:

- Incident report
- Event notification forms
- Incident investigation and analysis report
- Control center logs
- Station logs
- Photographs and notes
- Work orders, permits, and lockout tags
- Recommendations for corrective actions and training for personnel
- Documents showing corrective actions taken and training completed

Pipeline companies can take a number of actions to prevent some types of AOCs. Following are examples of some of the things pipeline companies may do to prevent AOCS:

- Replace or repair equipment
- Train or retrain employees
- Install additional backup systems
- Examine work procedures

2.2.0 Required Notifications and Reports

For some types of AOCs, an authorized employee must call the National Response Center (NRC) in Washington, D.C. A federal law describes what types of emergencies to report and how to report them (*49 CFR 191.3*). This law applies to natural, flammable, toxic, and corrosive gases. According to this regulation, AOCs that must be reported to the NRC include the following:

- A release of gas from a pipeline
- A release of liquefied natural gas
- Death
- Hospitalization
- Property damage, including cost of gas lost, of $50,000 or more
- Emergency shut in
- Any event that in the judgment of the pipeline company should be reported

Safety-Related Condition Report

Date: ________________

Government Letter #: ________

APSC FILE #: ______________

Information Resources Manager
Office of Pipeline Safety
Research and Special Programs Administration
U.S. Department of Transportation
Room 8417
400 Seventh Street SW
Washington, D.C. 20590

Safety-Related Condition Report

49 CFR 191.23
Subject:

Operator:

Date of Report:

Person Submitting Report: Senior Vice President
Business Unit

Person Determining Condition:

Name:

Title:

Telephone:

Date Condition Discovered:

Date Condition Determined:

Location of Condition:

Description of Condition:

Corrective Action:

Sincerely,

Sr. Vice President, Business Unit

107F01.EPS

Figure 1 ◆ Sample safety-related condition report.

NOTICE: This report is required by 49 CFR Part 191. Failure to report can result in a civil penalty not to exceed $25,000 for each violation for each day that such violation persists except that the maximum civil penalty shall not exceed $500,000 as provided in 49 USC 1678.

Form Approved
OMB No. 2137-0522

U.S. Department of Transportation
Research and Special Programs Administration

INCIDENT REPORT - GAS TRANSMISSION AND GATHERING SYSTEMS

Report Date ______________

No. ______________
(DOT Use Only)

INSTRUCTIONS

Important: *Please read the separate instructions for completing this form before you begin. They clarify the information requested and provide specific examples. If you do not have a copy of the instructions, you can obtain one from the Office Of Pipeline Safety Web Page at* http://ops.dot.gov.

PART A – GENERAL REPORT INFORMATION Check one: **Original Report** **Supplemental Report** **Final Report**

Operator Name and Address

a. Operator's 5-digit Identification Number (when known) /__/__/__/__/__/

b. If Operator does not own the pipeline, enter Owner's 5-digit Identification Number (when known) /__/__/__/__/__/

c. Name of Operator ______________

d. Operator street address ______________

e. Operator address ______________
City, County or Parrish, State and Zip Code

2. Time and date of the incident

/__/__/__/__/ hr. /__/__/ month /__/__/ day /__/__/ year

3. Location of incident

a. ______________
Nearest street or road

b. ______________
City and County or Parrish

c. ______________
State and Zip Code

d. Mile Post/Valve Station ______________

e. Survey Station No. ______________

f. Latitude: __________ Longitude: __________
(if not available, see instructions for how to provide specific location)

g. Class location description
Onshore: Class 1 Class 2 Class 3 Class 4
Offshore: Class 1 *(complete rest of this item)*
Area ______________ Block # ______________
State /__/__/ or Outer Continental Shelf

h. Incident on Federal Land other than Outer Continental Shelf
Yes No

i. Is pipeline Interstate Yes No

4. Type of leak or rupture

Leak: Pinhole Connection Failure *(complete sec. F5)*
Puncture, diameter *(inches)* ______________

Rupture: Circumferential – Separation
Longitudinal
– Tear/Crack, length *(inches)* _
- Propagation Length, total, both sides *(feet)* ______

N/A

Other: __________

5. Consequences *(check and complete all that apply)*

a. Fatality Total number of people: /__/__/__/
Employees: /__/__/__/ *General Public:* /__/__/__/
Non-employee Contractors: /__/__/__/

b. Injury requiring inpatient hospitalization Total number of people: /__/__/__/
Employees: /__/__/__/ *General Public:* /__/__/__/
Non-employee Contractors: /__/__/__/

c. Property damage/loss *(estimated)* Total $__________
Gas loss $__________ Operator damage $__________
Public/private property damage $__________

d. Release Occurred in a 'High Consequence Area'

e. Gas ignited – No explosion f. Explosion

g. Evacuation *(general public only)* /__/__/__/__/ people
Reason for Evacuation:
Emergency worker or public official ordered, precautionary
Threat to the public Company policy

6. Elapsed time until area was made safe:
/__/__/ hr. /__/__/ min.

7. Telephone Report
/__/__/__/__/__/__/ NRC Report Number /__/__/ month /__/__/ day /__/__/ year

8. a. Estimated pressure at point and time of incident:
______________ *PSIG*

b. Max. allowable operating pressure *(MAOP)*: __________ *PSIG*

c. MAOP established by 49 CFR section:
192.619 (a)(1) 192. 619 (a)(2) 192. 619 (a)(3)
192.619 (a)(4) 192. 619 (c)

d. Did an overpressurization occur relating to the incident? Yes No

PART B – PREPARER AND AUTHORIZED SIGNATURE

______________ (type or print) Preparer's Name and Title

______________ Area Code and Telephone Number

______________ Preparer's E-mail Address

______________ Area Code and Facsimile Number

______________ Authorized Signature ______________ (type or print) Name and Title ______ Date ______________ Area Code and Telephone Number

Form RSPA F 7100.2 (01-2002) ***Reproduction of this form is permitted*** *Page 1 of 3*

107F02A.EPS

Figure 2 ◆ Sample incident report (1 of 3).

PART C - ORIGIN OF THE INCIDENT

1. Incident occurred on
   - Transmission System
   - Gathering System
   - Transmission Line of Distribution System
2. Failure occurred on
   - Body of pipe
   - Pipe Seam
   - Joint
   - Component
   - Other: ______________________________
3. Material involved *(pipe, fitting, or other component)*
   - Steel
   - Plastic (If plastic, complete all items that apply in a-c)
   - Plastic failure was: a.ductile b.brittle c.joint failure
   - Material other than plastic or steel: ____________________
4. Part of system involved in incident
   - Pipeline
   - Regulator/Metering System
   - Compressor Station
   - Other: ________________
5. Year the pipe or component which failed was installed: / / / / /

PART D – MATERIAL SPECIFICATION (if applicable)

1. Nominal pipe size *(NPS)* / / / / / in.
2. Wall thickness / / / / / in.
3. Specification ______________ SMYS / / / / / / /
4. Seam type ____________________________
5. Valve type ____________________________
6. Pipe or valve manufactured by ______________________________________ in year / / / / /

PART E – ENVIRONMENT

1. Area of incident
   - Under pavement
   - Under ground
   - Inside/under building
   - In open ditch
   - Above ground
   - Under water
   - Other: ____________
2. Depth of cover: ______________ inches

PART F – APPARENT CAUSE

Important: *There are 25 numbered causes in this section. Check the box to the left of the* ***primary*** *cause of the incident. Check one circle in each of the supplemental items to the right of or below the cause you indicate. See the instructions for this form for guidance.*

F1 – CORROSION

1. External Corrosion
2. Internal Corrosion

If either F1 (1) External Corrosion, or F1 (2) Internal Corrosion is checked, complete all subparts a – e.

a. Pipe Coating
   - Bare
   - Coated

b. Visual Examination
   - Localized Pitting
   - General Corrosion
   - Other: ______________

c. Cause of Corrosion
   - Galvanic
   - Stray Current
   - Improper Cathodic Protection
   - Microbiological
   - Stress Corrosion Cracking
   - Other: ______________

d. Was corroded part of pipeline considered to be under cathodic protection prior to discovering incident?
   No Yes, Year Protection Started: / / / / /

e. Was pipe previously damaged in the area of corrosion?
   No Yes, How long prior to incident: / / / / *years* / / / *months*

F2 – NATURAL FORCES

3.	Earth Movement	Earthquake	Subsidence	Landslide	Other: ______________	
4.	Lightning					
5.	Heavy Rains/Floods	Washouts	Flotation	Mudslide	Scouring	Other: ________
6.	Temperature	Thermal stress	Frost heave	Frozen components		Other: ________
7.	High Winds					

F3 - EXCAVATION

8. Operator Excavation Damage *(including their contractors)* / Not Third Party
9. Third Party Excavation Damage *(complete a-d)*
   - a. Excavator group
     General Public Government Excavator other than Operator/subcontractor
   - b. Type: Road Work Pipeline Water Electric Sewer Phone/Cable Landowner Railroad
     Other: ____________________
   - c. Did operator get prior notification of excavation activity?
     No Yes: Date received: / / / mo. / / / day / / / yr.
     Notification received from: One Call System Excavator Contractor Landowner
   - d. Was pipeline marked?
     No Yes *(If Yes, check applicable items i – iv)*
     - i. Temporary markings: Flags Stakes Paint
     - ii. Permanent markings: Yes No
     - iii. Marks were *(check one)* Accurate Not Accurate
     - iv. Were marks made within required time? Yes No

F4 – OTHER OUTSIDE FORCE DAMAGE

10. Fire/Explosion as primary cause of failure Fire/Explosion cause: Man made Natural
11. Car, truck or other vehicle not relating to excavation activity damaging pipe
12. Rupture of Previously Damaged Pipe
13. Vandalism

Form RSPA F 7100.2 (01-2002) *Page 2 of 3*

107F02B.EPS

Figure 2 ◆ Sample incident report (2 of 3).

F5 – MATERIAL AND WELDS

Material

14. Body of Pipe — Dent — Gouge — Wrinkle Bend — Arc Burn — Other: ________
15. Component — Valve — Fitting — Vessel — Extruded Outlet — Other: ________
16. Joint — Gasket — O-Ring — Threads — Other: ________

Weld

17. Butt — Pipe — Fabrication — Other: ________
18. Fillet — Branch — Hot Tap — Fitting — Repair Sleeve — Other: ________
19. Pipe Seam — LF ERW — DSAW — Seamless — Flash Weld
    HF ERW — SAW — Spiral — Other: ________

*Complete a-g if you indicate **any** cause in part F5.*

a. Type of failure:

Construction Defect — Poor Workmanship — Procedure not followed — Poor Construction Procedures

Material Defect

b. Was failure due to pipe damage sustained in transportation to the construction or fabrication site? Yes No

c. Was part which leaked pressure tested before incident occurred? Yes, *complete d-g* No

d. Date of test: /__/__/ mo. /__/__/ day /__/__/ yr.

e. Test medium: Water — Natural Gas — Inert Gas — Other: ________

f. Time held at test pressure: /__/__/ hr.

g. Estimated test pressure at point of incident: ________ *PSIG*

F6 – EQUIPMENT AND OPERATIONS

20. Malfunction of Control/Relief Equipment — Valve — Instrumentation — Pressure Regulator — Other: ________
21. Threads Stripped, Broken Pipe Coupling — Nipples — Valve Threads — Mechanical Couplings — Other: ________
22. Ruptured or Leaking Seal/Pump Packing

23. Incorrect Operation

a. Type: Inadequate Procedures — Inadequate Safety Practices — Failure to Follow Procedures — Other: ________

b. Number of employees involved who failed post-incident drug test: /__/__/__/ Alcohol test: /__/__/__/

c. Were most senior employee(s) involved qualified? Yes No — d. Hours on duty: /__/__/

F7 – OTHER

24. Miscellaneous, *describe:* ________
25. Unknown

Investigation Complete — Still Under Investigation *(submit a supplemental report when investigation is complete)*

PART G – NARRATIVE DESCRIPTION OF FACTORS CONTRIBUTING TO THE EVENT *(Attach additional sheets as necessary)*

Form RSPA F 7100.2 (01-2002) — *Page 3 of 3*

107F02C.EPS

Figure 2 ◆ Sample incident report (3 of 3).

MARZETTA SIGNS 107F03.TIF

Figure 3 ◆ Lockout tag.

If any employees are injured, the pipeline company must also send a report to the Occupational Safety and Health Administration (OSHA) in Washington, D.C. If an employee dies, the company must immediately notify both the NRC and OSHA. The company must then send a written report to the NRC as soon as possible but no longer than 30 days from the day of the incident.

Some information, such as the amount of property damage and the cost of gas lost, may be estimates in the first report made by the pipeline company. The company must do a follow-up report when this information is finalized.

3.0.0 ◆ ABNORMAL FACILITY CONDITIONS

Any problem with the physical structure of the pipeline itself is defined as an abnormal facility condition. Following are examples of these problems:

- Uncovered pipeline that should be covered
- Damaged pipeline
- Missing or damaged pipeline supports
- Corrosion
- Coating separating from the pipeline (**disbondment**)
- Equipment on the pipeline right-of-way
- Missing signs and markers

Some reports may require immediate action. Others may require scheduling repairs at a later date. In all cases, the appropriate personnel must be notified so that repairs can be made safely and in a timely fashion.

3.1.0 Identification

Usually, work crews or line patrols report abnormal facility conditions. However, a report may come as a result of a call made to the One-Call notification system or from a person who calls the pipeline company directly. When a report comes in, it is important to get complete information. This information allows work crews to quickly find the problem and address it. Many pipeline companies use a standard form with blank lines for the following information:

- Caller's name and telephone number
- Date and time of the call
- Exact location of the problem
- Nearby landmarks that can help repair crews pinpoint the location
- Conditions observed at the site

Damage to pipeline supports (see *Figure 4*) is a serious problem. However, if the damage is not too bad, the company may decide not to shut in the pipeline segment. The decision to continue operations is a judgment call made by management and qualified inspection personnel. If the pipeline segment must be shut in, the control center notifies affected personnel. Whether the segment is shut in or not, a damage report must be submitted so that the maintenance supervisor can schedule repair work.

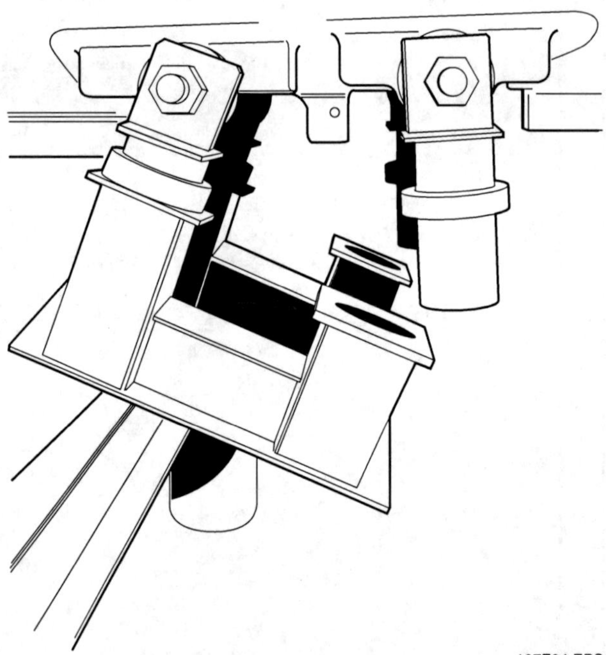

107F04.EPS

Figure 4 ◆ Damaged pipeline support.

3.2.0 Response

Response to damage reports must be quick but not hasty. Keep in mind that a caller's description may not be entirely accurate. The problem may be better or worse than described. In rare cases, a call might be a hoax. Therefore, when control center personnel receive a damage report, they must first confirm the report. They do this by checking readings on their monitors or by checking with field personnel. When a report is confirmed, field personnel should take the following steps:

Step 1 Investigate and evaluate the report.

Step 2 Carry out company safety procedures to protect people, property, and the environment if necessary.

Step 3 Notify control center personnel so that repairs can be scheduled.

Step 4 Complete the required reports.

4.0.0 ◆ ACTIVATION OF A SAFETY DEVICE

Safety devices are designed to sense hazardous conditions and alert personnel. These conditions include changes in pressure that might exceed pipeline design limits and releases of harmful gases. Safety devices include pressure relief valves and hazardous air alarms.

4.1.0 Identification

When a safety device activates, it sounds an alarm. The device may ring loudly, beep continuously, or flash a light. A sensor on the device also sends a signal to the control center, where it shows up as an alert on the control center console. You must never ignore a safety alarm. In addition, you must never turn off a safety alarm until the problem has been checked and cleared by authorized personnel.

4.2.0 Response

Personnel must respond to safety device alarms quickly and calmly. To respond to an alarm, personnel should follow these steps:

Step 1 Ensure that the alarm is signaling an AOC.

Step 2 Notify affected personnel.

Step 3 Carry out company safety procedures to protect people, property, and the environment.

Step 4 Troubleshoot the problem and make necessary repairs if qualified to do so.

Step 5 Reset and test the safety device alarm.

Step 6 Complete the required reports.

It is important to ensure that the alarm is not a false alarm. In case of a false alarm, the control center should notify affected employees and have the alarm equipment checked. Generally, no further action is required when an alarm is false.

When a hazardous gases alarm sounds, the response is somewhat different. In this case, personnel wearing the appropriate personal protective equipment (PPE) must follow these steps:

Step 1 Clear the area.

Step 2 Manually open any vents that do not automatically open to allow vapors to dissipate.

Step 3 Take readings to ensure that the air is safe to breathe.

Step 5 Reset the safety device and confirm that it is working properly.

Step 6 Complete the required reports.

NOTE

During an emergency, employees must focus on three key areas: ensuring employee safety, reducing the damage to equipment, and protecting the environment. The general rule is "people first." If possible, employees must try to keep a problem from turning into a disaster. However, all responses must also be made safely. Employees must wear the appropriate PPE. If they are told to clear an area, they must leave immediately.

5.0.0 ◆ COMMUNICATIONS FAILURE

Communications can be divided into two areas: systems communications and employee communications. Systems communications refers to the computer equipment located in the control center and on some pipeline equipment. Employee communications refers to the use of telephones, cell phones, pagers, or radios.

5.1.0 Systems Failure

Personnel in the control center use the Supervisory Control and Data Acquisition (SCADA) system to manage pipeline operations. They receive signals from and send signals to pipeline equipment that may be many miles away. The SCADA system allows them to monitor pressure, volume, and temperature and keep the gas flowing safely and on schedule. If the system fails, control center personnel lose the ability to communicate with equipment on the line. Therefore, most systems are set up to switch to a **redundant system**—a backup that comes online if the main system fails.

5.1.1 Identification

Most failures of communications systems show up as warnings or messages on the control center console. A systems failure may include any of the following events:

- Random facility outages
- Partial facility outages
- Complete facility outages
- Loss of the SCADA system
- Loss of satellite communications
- Loss of microwave signals

Mechanical failures are related to systems failures. For example, the SCADA system may be able to send a signal to a piece of equipment, but the equipment does not respond to the signal. In some cases, the problem is a result of a programming error. In other cases, equipment may not work because of a mechanical, electrical, or pneumatic problem. Examples include frozen valves, electrical component burnouts, and compressor breakdown.

5.1.2 Response

The control center must notify the appropriate personnel so that they can isolate, investigate, and address the problem. The appropriate responses may include some or all of the following activities:

- Switching to alternate or backup systems
- Shutting in a pipeline segment if necessary
- Determining the scope of the problem
- Taking corrective actions to restore the system
- Estimating how long the main system will be down
- Notifying affected personnel about how long repairs will take
- Completing the required reports

If the problem is with the SCADA system, control center personnel may have to call the SCADA support group or contractor. If equipment cannot be operated from the control center, manual operation in the field may be required.

Personnel in the affected area must be notified immediately. Generally, the pipeline can continue to operate while repairs are made.

5.2.0 Employee Communications Failure

Pipeline employees use radios, telephones, cell phones, and pagers to communicate with one another on the job. These tools usually are reliable, but breakdowns can occur. Employees must keep their equipment in good working order. They should also test their equipment from time to time to ensure that it is working properly.

> **WARNING!**
> Electronic equipment must not be used in the presence of hazardous gases unless it is classified as **intrinsically safe/explosion proof.** Failure to use this type of equipment in a hazardous atmosphere may result in an explosion.

> **NOTE**
> Intrinsically safe/explosion-proof electrical equipment is designed so that a spark cannot ignite the air around it. This type of equipment must be certified to meet the standards of the *National Electric Code®*. It is made to meet the following specifications:
> - Outside air cannot penetrate the casing.
> - The casing must be strong enough to contain an explosion inside the equipment without igniting the surrounding air.
> - The interior must be filled with nonflammable powder or inert gas to prevent ignition.

5.2.1 Identification

A breakdown in employee communications equipment is easy to identify. Workers cannot hear messages or make calls. Some equipment beeps or flashes a liquid crystal display to alert users to a problem.

5.2.2 Response

Workers should first check their equipment to see if the problem is a simple one that can be quickly and easily fixed. For example, some telephones have a volume control that can accidentally be turned all the way down. The telephone may seem to be dead, but a simple adjustment to the dial fixes the problem. Other simple solutions include replacing a battery or recharging a unit.

Some communications problems are more serious. Weather or problems at the local utility company can knock out communications completely. When there is a major failure in employee communications, workers must follow these steps:

Step 1 Switch to alternate equipment if necessary.

Step 2 Notify the appropriate personnel per company policy.

Step 3 Complete any required reports.

If communications are interrupted for long periods, it may not be possible to ask for instructions from a remote location. Training before an emergency occurs is the best way to ensure that you respond correctly. You must, however, do everything that you can to reestablish communications as soon as possible.

6.0.0 ◆ POWER INTERRUPTIONS

Electric power is essential to pipeline operations. Nearly every part of the pipeline depends on electricity in some way. Power interruptions that result from bad weather or a problem at the local utility company are not uncommon. If power goes off, electronic indicators turn off and some compressors and remote control valves might not operate. In addition, lights, heat, and air conditioning may fail. An area-wide power failure may also result in a loss of telephone communications.

6.1.0 Identification

Some pipeline companies have an uninterruptible power system (UPS) that keeps the pipeline running in case of a power failure. Instead of electricity, these systems run on engines fueled with gasoline, diesel, or natural gas. UPS systems turn on if electrical power is off for more than a few seconds. When power fails, an alternating current (AC) power failure alarm is sent to the control center. To ensure that the backup system has turned on, control center personnel check to see whether they are receiving updated information from the pipeline. The backup system allows operations to continue until AC power can be restored.

Some equipment is protected from power failure by batteries that supply temporary low voltage. These batteries run on direct current (DC). The following types of equipment can operate on DC power:

- Fire detection equipment and alarms
- Fire suppression systems
- Emergency lighting systems
- Alarm panels
- Control modules
- Process control panels
- Specialized compressor units

> **NOTE**
> AC power reverses its direction at regular intervals. Equipment plugged into electric sockets uses AC power. DC power flows in one direction. Batteries supply DC power.

6.2.0 Response

The appropriate response to a power failure depends on the scope of the problem. For example, if power fails at only one station, control center personnel can make adjustments to keep the pipeline operating. However, if that station is needed to keep pressure below MAOP, personnel may have to shut in a pipeline segment. The general response to a power interruption includes these steps:

Step 1 Follow company troubleshooting procedures for power failures.

Step 2 Ensure that the UPS system has activated.

Step 3 Switch to an alternate control system or power source if necessary.

Step 4 Notify affected personnel.

Step 5 Complete the required reports.

7.0.0 ◆ FIRE, EXPLOSION, OR NATURAL DISASTER

The pipeline company must have a training program in place so that employees know what to do in case of fire, explosion, or natural disaster. Many companies conduct regular drills so that employees can practice their responses to such events. Drills help ensure that employees are able to calmly handle emergencies safely. Each company has its own procedures for emergencies. In general, employees should know the following information:

- The location of fire suppression equipment and how to operate it
- The location of first aid equipment and how to use it

> **NOTE**
> It is important to provide first aid correctly. Only employees trained and certified in first aid procedures may provide first aid.

- The process for shutting down equipment manually if communication with the control center is lost
- The company's evacuation plan
- The procedures for contacting local emergency response agencies if necessary

Depending on the problem, pipeline employees may be able to safely handle an emergency without calling for outside help. You must follow your company's policy for contacting local emergency response agencies.

7.1.0 Fire

Detection devices installed along the pipeline and in pipeline facility buildings alert employees in case of a fire. These devices send an alarm when their sensors detect heat or flame. A thermal detector is a heat-sensitive unit. It is fitted with electrical contacts that stay open as long as conditions are normal. When the unit senses rapidly rising temperatures, the contacts start to close. When the contacts connect, the unit sends an alarm to the control center and the fire control panel in the workstation.

A flame detector is a light-sensitive unit. It is fitted with an ultraviolet detector that turns on an alarm when the unit senses light from a flame. The unit is sensitive enough to detect a flame even in the earliest stages of a fire. Some flame detectors are connected to a firefighting system that immediately dumps fire suppression material into the area. Employees can also dump this material over a fire manually by using pull switches.

> **CAUTION**
> Before using any fire suppression material, you must know how to handle it. Be sure to read the material safety data sheet (MSDS) for information about how to use the material safely.

7.1.1 Response

Control center personnel must ensure that the fire alarm is not false. If there is no fire, the following steps may be taken:

Step 1 Notify affected personnel.

Step 2 Check or test alarms to ensure that they are operating properly.

Step 3 Reset alarms or ensure that they have automatically reset.

Response to a fire must be quick, automatic, and calm. The goal is to limit the damage as much as possible to protect lives and property. The focus should be on people first, property second. Injured employees must receive medical attention as soon as possible. Once the fire is out, the air is cleared, and any injured employees have received medical attention, the following actions are taken:

- Alarms are reset and tested.
- Equipment damage is assessed and reported to the appropriate personnel.
- Work orders are issued and required permits are obtained for repairs.
- Employee performance is evaluated and corrective action is determined.
- Injuries or deaths are reported to the appropriate regulatory agencies if required.

The evaluation of employee performance during a fire may reveal problems in employee response. Management generally recommends retraining so that employees can improve their responses to a fire.

Fire extinguishers, which may also be used to put out a fire, may be located in many work areas. Fire extinguishers are designed for specific types of fires. It is important to know the extinguisher's fire class rating before using it. The fire class rating is printed at the bottom of the label on a fire extinguisher. There are four fire class ratings (see *Table 2*).

7.2.0 Explosion

Pipeline companies install a number of safety devices to warn personnel about conditions that could cause explosions. However, companies cannot control every circumstance, such as intentional damage, that could lead to an explosion on the pipeline. An explosion on a transmission or distribution pipeline registers in the control center

Table 2 Fire Class Ratings

Fire Class	Materials and Proper Fire Extinguisher
Class A fire	Involves ordinary combustibles, such as wood or paper. This type of fire is suppressed by cooling the fire. Class A extinguishers contain water.
Class B fire	Involves grease, flammable liquids, and gases. Class B extinguishers contain carbon dioxide or another material that smothers the fire by removing oxygen.
Class C fire	Involves electrical equipment. Class C extinguishers are designed to protect the firefighter from electric shock. This type of extinguisher smothers the fire.
Class D fire	Involves metals. Class D extinguishers contain a powder that either forms a crust around the burning metal or gives off gases that prevent oxygen from reaching the fire. Some metals may keep burning even after they have been coated with powder from a Class D extinguisher. The best way to fight these fires is to keep using the extinguisher so that the fire does not spread.

as an unexplained change in pressure. An explosion may occur in any enclosed space where natural gas can accumulate to dangerous levels. Examples of such spaces include deep pipeline trenches and facility buildings.

> **NOTE**
>
> A natural gas explosion in a home or office may occur when natural gas leaks into an enclosed space. The odorant added to the gas (which smells like rotten eggs) usually alerts people to the problem. However, the gas can sometimes accumulate undetected. For example, the building may be vacant or the odors may be too far away from people inside the building. A spark from electrical equipment can then cause the gas to explode.

7.2.1 Response

A response to an explosion must be quick, decisive, and calm. If the main-line valves do not close automatically, personnel must initiate actions to ensure that they are closed. Each company has specific procedures for responding to an explosion. In addition, the correct order of the steps to take in response to an explosion may vary from company to company. In general, these are the appropriate responses to an explosion:

- Evaluate and assess the situation
- Carry out company safety procedures to protect people, property, and the environment
- Call local emergency response agencies if necessary
- Confirm the location of all employees near the explosion location
- Confirm the location of any members of the public near the explosion location
- Conduct an incident investigation
- Call local, state, or federal law enforcement if the cause indicates criminal or terrorist activity
- Complete the required reports

7.3.0 Natural Disaster

All pipeline companies watch the weather closely and pay special attention to weather that can develop into a natural disaster. Control center personnel monitor for earthquakes with seismic indicators. In addition, they may obtain real-time information on the Internet. Up-to-date earthquake information is available from the National Earthquake Information Center, which is located in Denver, Colorado. Personnel may track weather alerts and warnings through the National Weather Service Web site. In case of a power failure, personnel can continue to receive information through the National Weather Service's National Weather Radio. A list of weather-related organizations is included in *Appendix B.*

7.3.1 Response

Thanks to the agencies that track weather and other conditions, pipeline companies have time to prepare for a natural disaster, such as an earthquake, hurricane, or tornado. Most of the pipeline's ability to withstand extreme weather is already present in its design and installation. A great deal of care and thought goes into pipeline design. In addition, pipeline companies perform thorough surveys and tests and must meet federal safety standards when installing pipelines.

Each company has specific procedures for dealing with natural disasters; these reflect local conditions. For example, a hurricane is most likely to occur in Florida or the Gulf states. Therefore, companies located in those areas include hurricane response in their procedures. Following are examples of appropriate responses to weather-related problems:

- Monitoring developing weather conditions such as tropical storms that may become hurricanes
- Carrying out company safety procedures for protecting people, property, and the environment

> **NOTE**
>
> During a natural disaster, the goal is to restore interrupted service as quickly as possible. Crews may work around the clock, sometimes in dangerous conditions, to achieve this goal. However, the focus is always on safety.

- Adjusting work schedules as needed to respond to emergency conditions
- Following company procedures for responding to fire, explosion, or damage resulting from natural disasters
- Setting up a command post if conditions warrant such a step
- Following company procedures for administering first aid if required
- Working with local emergency response agencies if required
- Conducting the required investigations
- Completing the required reports

7.4.0 Evacuations

Evacuations are necessary when attempts to control a problem fail and there is an impending threat to lives. Evacuations in the pipeline industry can be divided into two areas: evacuation of the pipeline facility and evacuation of the public. Some emergencies require evacuating both the facility and the public.

7.4.1 Evacuation of the Pipeline Facility

Certain conditions may require an immediate evacuation of the pipeline facility. The decision to evacuate is based on a number of factors and is made on a case-by-case basis. For example, a small fire that is quickly put out usually does not require an evacuation. Release of a hazardous gas may require that all employees immediately clear the area. In general, the following items are included in an evacuation plan for employees:

- Evacuation maps posted in every area and department of the pipeline facility
- Designated safe areas that employees must report to if evacuated
- Warning signs to post on doors of areas that are considered too hazardous to enter
- A procedure for checking on the whereabouts of each employee
- A procedure for ensuring that the area is safe for employees to re-enter

When the pipeline facility is evacuated, it is important to account for every employee as soon as possible. Certain employees may act as emergency coordinators in safe areas. Their job is to conduct a roll call to confirm that all employees are safe. If any employees cannot be accounted for, authorized personnel must locate and, if necessary, rescue them. Rescue teams must have complete information before entering a hazardous environment. They must know whether any employees are trapped and where they may be located. Emergency evacuations can result in a certain amount of confusion. Employees must help eliminate confusion by doing the following:

- Reporting to their designated safe area and identifying themselves to the safety coordinator
- Providing accurate information about any employees who may be trapped in the facility
- Staying away from the danger area unless assigned to a rescue team
- Keeping the lines of communication clear

7.4.2 Evacuation of the General Public

Some emergencies may require an evacuation of the public located near the pipeline. The pipeline company is not authorized to order a general evacuation. Only the local emergency response authorities may do so. However, the pipeline company may participate in managing the evacuation. In case of a major disaster, companies work with the local authorities to accomplish the following tasks:

- Determine whether a general evacuation is required
- Provide information about the AOC
- Provide contact information for people located near the pipeline
- Provide personnel to help staff a command center if requested

NOTE

Local emergency planning committees (LEPCs) exist in every county in the United States. Law enforcement agencies, firefighting companies, local government agencies, hospitals, companies, and citizen's groups make up the memberships of LEPCs. Committee members meet on a regular basis to plan for emergencies. Planning for a general evacuation is an example of just one of the projects handled by an LEPC.

Federal law requires pipeline companies to have a damage prevention program (*49 CFR 192.614*). This program must include a plan for alerting everyone in the vicinity of the pipeline in case of an emergency. The term *vicinity* is not defined in the law. Generally, the company decides how large the area should be with help from the appropriate regulatory officials.

In major disasters, the local authorities may set up a command center in a safe location. In command centers, the communications systems are set up to run off batteries or generators. Communications are therefore able to continue even during an area-wide power outage. Pipeline personnel may be asked to serve as subject matter experts at the command center. Subject matter experts can give advice and information to emergency response teams to help them do their jobs safely and more efficiently.

8.0.0 ◆ PIPELINE SYSTEM DAMAGE

Damage from outside sources may affect the pipeline or its components. This type of damage can be divided into two areas: accidental and intentional.

The location of the pipeline is clearly marked to prevent accidents. However, construction work may cover markings, and a worker may strike the line without realizing it. An excavator may lose control of a bulldozer or other heavy equipment and hit the line. In some parts of the country, hunters have accidentally shot pipelines, and ships' anchors may hit offshore pipelines.

Intentional damage may be the result of criminal or terrorist activity. This type of damage may require the involvement of local, state, and federal law enforcement officials.

8.1.0 Identification

A person or company who causes accidental damage to the pipeline is required by law to notify the One-Call notification system (*49 CFR 190.229*). The One-Call operator alerts the pipeline company. Damage also shows up on the control center console as an unexplained change in operating conditions.

8.2.0 Response

The control center confirms the damage report and notifies affected personnel. A crew is then sent to investigate. Each company has specific procedures for responding to this type of damage. Therefore, you must always follow your company's policy. In general, the correct response to accidental or intentional damage includes the following actions:

- Investigating and evaluating the problem
- Lessening damage to the line
- Carrying out company safety procedures to protect people, property, and the environment
- Notifying the appropriate company personnel
- Following company procedures for contacting local emergency response agencies if necessary
- Following company procedures for contacting local, state, or federal law enforcement agencies if necessary
- Notifying repair crews and issuing work orders
- Ensuring that all repairs are completed
- Completing the required reports

Once the pipeline has resumed normal operations, the pipeline company conducts an internal investigation. The company's investigation focuses on evaluating employee performance, control systems, and mechanical equipment. If the damage was intentional, regulatory and law enforcement agencies conduct a criminal investigation. These agencies may ask for results of the pipeline company's own investigation.

9.0.0 ◆ HAZARDOUS GASES

Natural gas does not explode under normal circumstances. For an explosion to take place, there must be a source of ignition, and the gas must make up between 5 and 15 percent of the air. Generally, when natural gas escapes into the air, its molecules dissipate or scatter. However, in enclosed areas, natural gas can accumulate, and if there is an ignition source, an explosion occurs.

9.1.0 Identification

Gas detection devices are used to sense when hazardous gases are in the air. These gases include hydrogen sulfide (H_2S), natural gas, and, in some cases, carbon dioxide (CO_2). Generally, natural gas is most hazardous in contained areas where it can build to a high concentration. Workers may use a combustible gas indicator or a flame ionization detection unit to detect hazardous gases. Workers must not rely on sense of smell to detect a hazardous gas for the following reasons:

- Some hazardous gases are odorless.
- A low concentration of some hazardous gases can deaden the sense of smell.
- The sense of smell may be deadened by a medical problem.

9.2.0 Response to a Natural Gas Leak

A response to a natural gas leak must be quick, decisive, and calm. If a natural gas leak is detected in a pipeline facility, trained personnel may perform some or all of the following actions:

- Carry out company safety procedures to protect people, property, and the environment.
- Isolate the leak if it is safe to do so.
- Manually open doors and windows to vent the gases and allow vapors to dissipate.
- Have the power company cut power if appropriate.

Of equal importance are actions personnel must avoid. When responding to a natural gas leak, personnel must take these precautions:

- Do not open switchgear doors.
- Do not operate electrical switches or equipment, including lights.
- Do not operate electrical equipment unless it is classified as intrinsically safe/explosion proof.
- Do not make telephone calls unless the telephone equipment is classified as intrinsically safe/explosion proof.

> **WARNING!**
> In the presence of a combustible gas, you must not operate electrical equipment that is not classified intrinsically safe/explosion proof. Failure to use the correct type of equipment may result in an explosion.

9.3.0 Response to Other Gas Leaks

Gases other than natural gas may leak into the pipeline facility. If hazardous gases such as H_2S or CO_2 are detected, personnel wearing the appropriate PPE must take the following steps:

Step 1 Clear the area.

Step 2 Vent the area.

Step 3 Prevent anyone who is not wearing the appropriate PPE from entering the area.

Step 4 Call for emergency rescue and medical help if required.

> **WARNING!**
> H_2S is a colorless gas that smells like rotten eggs. However, in fairly low concentrations, it can deaden the sense of smell. Exposure at a higher concentration can cause death. In addition, it is flammable and may cause a flash fire. When a gas detection unit shows that H_2S is present, you must wear the appropriate PPE or immediately leave the area.

> **WARNING!**
> CO_2 is a colorless, odorless gas used as a propellant for fire-extinguishing equipment. It is also a by-product of combustion, as well as incomplete burning. In addition, some pipeline operations pump carbon dioxide commercially. Heavier than air, CO_2 can collect in low-lying areas where it may displace oxygen. Exposure can result in death.

Only personnel trained in rescue operations may enter an area where a hazardous gas has been detected. These personnel must be trained and qualified to use a self-contained breathing apparatus (see *Figure 5*). Other workers must stay a safe distance away until the deadly air is cleared. Warning signs must be posted at the doors so that workers do not enter until the area has been inspected and tested.

NORTH SAFETY PRODUCTS 107F05.EPS

Figure 5 ◆ Self-contained breathing apparatus.

Only personnel trained and certified in first aid may provide medical aid to injured employees. The instinct to move an injured person is natural. However, unless it is absolutely necessary, it is best not to move an injured person until qualified personnel arrive.

An authorized supervisor must ensure that a release of a hazardous gas and any injuries or deaths resulting from the release are properly noted for company records and reports. The paperwork may include the following:

- An incident report
- An incident investigation and analysis report
- Event notification forms
- Control center logs
- Station logs
- Recommendations for corrective actions and employee training or retraining
- Records of corrective actions taken and employee training completed

10.0.0 ◆ UNEXPLAINED FLOW RATE

Flow rate is an important indicator of both normal and abnormal operating conditions. Minor changes in flow rates may occur naturally on the pipeline, but significant changes are considered AOCs.

10.1.0 Identification

Meters located along the pipeline at custody transfer stations monitor flow rates and send readings to the control center. When an unexpected change in flow rate occurs, an alarm alerts control center personnel. Sudden flow rate increases or decreases could point to a possible release. Personnel must be familiar enough with the pipeline system, equipment, product characteristics, and environmental factors to quickly identify any important changes in flow.

10.2.0 Response

Control center personnel must investigate flow rate alarms to determine what the problem is at the metering station. Personnel should take the following steps:

Step 1 Investigate and evaluate the problem to learn why the alarm sounded.

Step 2 Check flow rates upstream and downstream for comparison purposes.

Step 3 Check the status of nearby compressors and valves to determine if a problem with the equipment caused the flow rate change.

If a release has not occurred, operations personnel should adjust flow and pressure to achieve normal operations. However, they must also find out the cause of the alarm. It may have sounded as a result of a problem in the SCADA system. It may have sounded as a result of an interruption of data transmission caused by maintenance workers performing repairs. No matter what the reason, the problem must be corrected, and the alarm must be reset.

11.0.0 ◆ UNEXPLAINED PRESSURE CHANGES

Pressure is another important indicator of both normal and abnormal operating conditions. Minor changes in pressure may occur normally on the line. However, significant pressure changes are considered AOCs. Unexplained pressure changes include unexplained absences of, increases in, and decreases in pressure.

11.1.0 Identification

Pressure-reading devices located near main-line valves, compressor stations, and custody transfer stations monitor pressure and send readings to the control center. When an unexpected change in pressure occurs, an alarm alerts control center personnel. Sudden pressure increases or decreases might indicate a problem such as the failure of a valve or a transmitter. Control center personnel must evaluate and investigate unexplained changes in pressure.

11.2.0 Response

Control center personnel must investigate pressure alarms to locate the problem. Once the problem is located, personnel take the following steps:

Step 1 Investigate and evaluate the problem to determine the reason for the alarm.

Step 2 Check the pressure upstream and downstream for comparison purposes.

Step 3 Notify the appropriate personnel, and shut in the pipeline segment if necessary.

Operations personnel should adjust flow and pressure to achieve normal operations. However, they must also find out the cause of the alarm. For example, a valve or transmitter may have failed. Once the problem has been determined, the control center is notified so that corrective actions can be taken. Generally, the control center notifies the maintenance supervisor, who schedules necessary repairs.

Summary

Because AOCs are a common occurrence, pipeline companies have developed effective routines and policies that enable employees to respond quickly, safely, and effectively. Usually an AOC is a routine matter that can be handled safely by pipeline personnel. Sometimes, however, an AOC can result in an emergency situation that may involve working with outside agencies.

In this module, you learned about the different types of AOCs that can occur on the pipeline. You also learned about the importance of reporting AOCs to government agencies that oversee pipeline operations. During your career, you may have to respond to some or all of the AOCs outlined in this module. When responding to these AOCs, you must follow your company's policies and procedures. Your safety and the safety of your co-workers and the general public depend on it.

Review Questions

1. Companies submit information about an AOC to the Office of Pipeline Safety on _____.
   a. MSDS reports
   b. DOT reports
   c. safety-related condition reports
   d. station logs

2. Following some types of AOCs, the company may be required by law to _____.
   a. conduct an investigation
   b. conduct a training session
   c. discipline employees who have not followed company policy
   d. test all system communications

3. A _____ must be reported to the NRC.
   a. missing pipeline support
   b. valve communication problem
   c. release of gas
   d. control center console problem

4. When a safety device sounds an alarm, you must _____.
   a. turn it off
   b. never ignore it
   c. reset it
   d. call the local emergency response agency

5. Electronic equipment must not be used in the presence of hazardous gases unless it is classified as _____.
   a. NRC approved
   b. OSHA approved
   c. intrinsically safe/explosion proof
   d. mechanically safe/disruption proof

6. Some pipeline companies have an _____ that keeps the pipeline running in case of a power failure.
   a. operations override power source
   b. uninterruptible power system
   c. interface power system
   d. indirect power system

7. Before using a fire extinguisher, it is important to know its _____.
   a. fire class rating
   b. OSHA rating
   c. safety features
   d. capacity

8. When an emergency evacuation of the pipeline facility occurs, all of the following are correct responses except _____.
   a. reporting to a designated safe area and identifying yourself to the safety coordinator
   b. providing first aid
   c. staying away from the danger area unless assigned to a rescue team
   d. keeping the lines of communication clear

9. A person or company who causes accidental damage to the pipeline is required by law to notify _____.
   a. OSHA
   b. the NRC
   c. the local authorities
   d. the One-Call notification system

10. Only personnel trained in rescue operations and qualified to use a _____ may enter an area where a hazardous gas has been detected.
    a. combustible gas indicator
    b. flame ionization indicator
    c. self-contained breathing apparatus
    d. fire suppression kit

GLOSSARY

Trade Terms Introduced in This Module

Disbondment: Separation of the protective coating from the pipeline.

Intrinsically safe/explosion proof: Describes electrical equipment that is designed specifically for use in explosive or flammable locations so that a spark cannot ignite the air around it.

Redundant system: A system that duplicates the operations of an entire system so that, in the event of a failure, a complete backup system is available.

APPENDIX A

Organizations of Interest to the Pipeline Industry

American Gas Association (AGA)
400 N. Capitol Street, NW
Washington, DC 20001
www.aga.org

American Petroleum Institute (API)
1220 L Street, NW
Washington, DC 20005
http://api-ec.api.org

Interstate Natural Gas Association of America (INGAA)
10 G Street, NE
Suite 700
Washington, DC 20002
www.ingaa.org

National Institute for Occupational Safety and Health (NIOSH)
Hubert H. Humphrey Building
200 Independence Avenue, SW
Room 715H
Washington, DC 20201
www.cdc.gov/niosh

National Response Center (NRC)
c/o United States Coast Guard (G-OPF)
2100 2nd Street, SW
Room 2611
Washington, DC 20593-0001
www.nrc.uscg.mil

Occupational Safety and Health Administration (OSHA)
200 Constitution Avenue, NW
Washington, DC 20210
www.osha.gov

Office of Pipeline Safety
U.S. Department of Transportation
400 7th Street, SW
Room 7128
Washington, DC 20590-0001
http://ops.dot.gov

APPENDIX B

Weather and Natural Disaster Information Sources

Emergency Managers Weather Information Network
14th Street and Constitution Avenue, NW
Washington, DC 20230
http://iwin.nws.noaa.gov/emwin/index.htm

National Earthquake Information Center
United States Geological Survey
Box 25046, DFC, MS 967
Denver, CO 80225
http://neic.usgs.gov

National Hurricane Center
Tropical Prediction Center
11691 SW 17th Street
Miami, FL 33165-2149
www.nhc.noaa.gov

National Weather Service and National Weather Service Weather Radio
National Oceanographic and Atmospheric Administration (NOAA)
14th Street and Constitution Avenue, NW
Room 6013
Washington, DC 20230
www.nws.noaa.gov

Figure Credits

North Safety Products 107F05

CONTREN™ LEARNING SERIES — USER UPDATES

The NCCER makes every effort to keep these textbooks up-to-date and free of technical errors. We appreciate your help in this process. If you have an idea for improving this textbook, or if you find an error, a typographical mistake, or an inaccuracy in NCCER's Contren™ textbooks, please write us, using this form or a photocopy. Be sure to include the exact module number, page number, a detailed description, and the correction, if applicable. Your input will be brought to the attention of the Technical Review Committee. Thank you for your assistance.

Instructors – If you found that additional materials were necessary in order to teach this module effectively, please let us know so that we may include them in the Equipment/Materials list in the Instructor's Guide.

Write: Curriculum Revision and Development Department
National Center for Construction Education and Research
P.O. Box 141104, Gainesville, FL 32614-1104

Fax: 352-334-0932

E-mail: curriculum@nccer.org

Craft ______________________ Module Name ______________________

Copyright Date ______________ Module Number ______________ Page Number(s) ______________

Description

____

____

____

____

(Optional) Correction

____

____

____

(Optional) Your Name and Address

____

____

____

Index

Index